RADIO ASTRONOMY

F. Graham Smith was born in 1923 and educated at Epsom College, Rossall School, and Downing College, Cambridge. During the war he was involved in radar research and from 1947 to 1964 he researched in radio astronomy at the Cavendish Laboratory with Sir Martin Ryle. He has been Professor at Jodrell Bank since 1964 and from 1964 to 1970 was Secretary of the Royal Astronomical Society. In 1970 he was elected a Fellow of the Royal Society and an Honorary Fellow of Downing College. Professor Smith's publications cover a wide range of work on Galactic and extra-galactic radio astronomy, and on space research.

F. Graham Smith is now Professor of Radio Astronomy at the Nuffield Radio Astronomy Laboratories, University of Manchester.

GW00381819

F. Graham Smith

RADIO ASTRONOMY

REVISED EDITION

*Hardly do we divine the things that are on
earth, and the things that are close at hand
we find with labour, but the things that are
in the heavens, who ever yet traced out?*

WISDOM OF SOLOMON 9:16

PENGUIN BOOKS

Penguin Books Ltd, Harmondsworth, Middlesex, England
Penguin Books Inc., 7110 Ambassador Road, Baltimore, Maryland 21207, U.S.A.
Penguin Books Australia Ltd, Ringwood, Victoria, Australia
Penguin Books Canada Ltd, 41 Steelcase Road West, Markham, Ontario, Canada
Penguin Books (N.Z.) Ltd, 182–190 Wairau Road, Auckland 10, New Zealand

—

First published 1960
Second edition 1962
Third edition 1966
Fourth edition 1974

—

Copyright © F. Graham Smith, 1960, 1962, 1966, 1974

—

Made and printed in Great Britain
by Richard Clay (The Chaucer Press) Ltd,
Bungay, Suffolk
Set in Monotype Times

Contents

LIST OF PLATES 7

LIST OF FIGURES 9

INTRODUCTION 11

1 *The Origins of Radio Astronomy* 15

2 *Cosmic Radio Waves* 24

3 *The Radio Universe* 33

4 *The Sun and its Corona* 52

5 *Sunspots, Flares and the Earth* 68

6 *Supernovae and Radio Stars* 81

7 *Pulsars* 94

8 *Interstellar Gas* 108

9 *Molecules in Space* 125

10 *Galaxies* 135

11 *The Mysterious Quasars* 156

12 *Radio Cosmology* 164

13 *The Moon* 177

14 *The Planets* 192

15 *Meteors and Comets* 207

16 *Radio Telescopes* 219

17 *The World's Radio-Astronomical Observatories* 249

INDEX 267

List of Plates

I The Andromeda nebula (M31)

II The Whirlpool nebula (M51)

III The Virgo A radio galaxy (M87)

IV The quasar 3C 273

V A group of four extragalactic nebulae in Leo

VI The Milky Way

VII Orion nebula

VIII Crab nebula (M1, the Taurus A radio source)

IX The Crab nebula pulsar

X The supernova remnant Cassiopeia A

XI Cygnus A

XII IC 443

XIII The solar corona

XIV Sunspots, 24 May 1947

XV Arch prominence, 4 June 1946

XVI Comet Bennett, 4 April 1969

XVII Jupiter

XVIII Meteor echo

XIX The Mills Cross radio telescope

XX Effelsberg 100-m reflector telescope

XXI Jodrell Bank Mark IA radio telescope

XXII Radar measurements of the planet Mars

Plates obtained from the library of the Royal Astronomical Society are designated by RAS and PAL catalogue numbers.

List of Figures

1. The spectrum of electromagnetic waves — 27
2. Sketch of a spiral galaxy — 42
3. Graph showing the temperature of radio waves emitted by the sun — 54
4. Radio-brightness distribution across the sun at long wavelengths — 56
5. Limb-brightened distribution of solar radio emission at 3-mm wavelength — 57
6. Radio eclipse of 30 June 1954 — 58
7. Bending of 80-MHz radio waves by the solar corona — 64
8. Diffraction in the solar corona — 65
9. Map of solar radiation at 1-cm wavelength — 72
10. The magnetic field over a pair of sunspots — 73
11. A solar flare observed by X-rays, 3-cm radio, ultra-violet light, and as a disturbance of the terrestrial ionosphere — 75
12. The dynamic spectrum of a solar outburst — 76
13. The dynamic radio spectrum of a Type II solar outburst — 78
14. Solar radio outburst of 30 March 1969 — 78
15. Polarized radio waves from the Crab nebula — 85
16. Radio remnant of Tycho Brahe's supernova — 89
17. Radio pulses from two of the pulsars discovered at Cambridge — 95
18. The changing period of the Vela pulsar — 98
19. The pulsed emission from the Crab nebula pulsar at radio, optical and X-ray wavelengths — 101
20. Magnetic field round a rotating neutron star — 102
21. Integrated pulse shapes from different pulsars — 104
22. Galactic radio map at 150 MHz — 109
23. Radio emission from the Milky Way — 111
24. The radio sources at the centre of the galaxy — 114
25. The profile of the spectral line emitted by interstellar hydrogen — 118
26. Profiles of the 21 cm line radiation from hydrogen at various longitudes in the galactic plane — 119
27. Spiral arms in our galaxy — 120
28. Faraday rotation in a spiral arm — 122
29. Scintillation of radio waves from the pulsar PSR 0329+54 — 124
30. Hydroxyl spectral line emission from the W3 nebula — 128
31. Spectral line emission from water vapour in the Orion nebula — 130
32. Spectral lines from elements in the Orion nebula — 132

33. Rotation of the galaxy M31, the Andromeda nebula   138–9
34. Radio red-shift velocities compared with optical red-shift velocities   141
35. A radio map of the Cygnus A radio galaxy   145
36. Radio sources in the Perseus cluster   148
37. The structure of four radio galaxies   151–2
38. The 'double double' radio source 3C 147   153
39. Model of a radio galaxy   155
40. Occultation of the quasar 3C 273 by the moon, 5 August 1962   157
41. The evolution of an oscillating universe   166
42. Radio source statistics. Plot of number against intensity, showing an excess of weak sources   170
43. Distribution of the galaxies in an evolutionary universe   171
44. The counts of weak radio sources   172
45. The radio temperature of the moon through one lunar month   182
46. The reflection of a radar pulse by the moon   186
47. Radar map of the moon   188
48. The distance of the moon found by triangulation   190
49. Jupiter's radiation belt   201
50. Radio waves from Jupiter   204
51. The origins of meteor showers   209
52. The measured velocities of meteors   215
53. Radio-telescope mounts   226
54. Early interferometer recordings of two radio sources   230
55. The development of the unfilled aperture   234
56. Aperture synthesis   236
57. The effect of increasing angular resolution   238
58. The variation of receiver noise with frequency   241
59. A simple interferometer radio telescope   247

Introduction

Yet it is possible that some bodies, of a nature altogether new, and whose discovery may tend in future to disclose the most important secrets in the universe, may be concealed under the appearance of very minute single stars in no way distinguishable from others of a less interesting character, but by the test of careful and repeated observations.

THESE prophetic words were spoken by one of Britain's great astronomers, Sir William Herschel, in January 1820. They anticipate by 143 years one of radio astronomy's historic achievements, the discovery of the quasars in 1963. Radio waves themselves were not discovered until 1887, long after Herschel's death, when Heinrich Hertz made his famous experiment with spark gaps and coils of wire. Even the concept that the universe is composed of myriads of galaxies, each like our own Milky Way containing myriads of stars, and all partaking in a general expansion, only dates back to the work of George Ellery Hale in the 1920s. The history of modern astronomy in fact stretches back over only half a life-span, an era which has seen a transformation of our view of the universe on all scales, in size from the smallest compact stars to the largest cosmological scale, and in time from the rapidly flashing light of a pulsar to the aeons of time in which the universe has evolved since its explosive creation.

Radio astronomy has been deeply involved in the new explorations of the universe since the early 1950s. There is a birth-date for the subject in 1932, when cosmic radio noise was first detected by Karl Jansky, but until 1946 there was only one radio astronomer, the remarkable amateur Grote Reber. At that time the stimulus of wartime research led to the establishment of several research groups, notably in England and Australia, and a rapid growth began. The name 'radio astronomy' was first used when the optical astronomers realized the value of measuring radio waves as well as the light emitted by various kinds of celestial

11

object; in particular the radio galaxy Cygnus A was seen to be a much more prominent object in the radio spectrum than it was visually. Since that time astronomy has extended into other parts of the electromagnetic spectrum, including infra-red light, ultraviolet light, X-rays, and gamma-rays. All these need special techniques which may be as different from optical astronomy as were the original radio techniques, but the contributions of radio have become so much part of astronomy that it now needs little adaptation for an astrophysicist to accept observations from any part of the spectrum. Furthermore, many of the techniques of radio astronomy, relying heavily on electronics, have spread through the whole spectrum, stimulating new approaches to old observing problems.

Radio astronomy has now become so much a part of the fabric of modern astronomy that it is difficult to distinguish it clearly as a separate subject. There are, however, special qualities of radio waves which make them especially suitable for the exploration of parts of the universe such as the gas clouds in our galaxy, the atmospheres of certain unusual stars such as the pulsars, and the explosive gas clouds of extragalactic objects such as the quasars. These qualities are discussed in the early chapters of this book, which attempts to place radio astronomy in the perspective of our present knowledge of the whole universe.

The more detailed chapters, which lead out from the sun, through the Milky Way, to radio galaxies, quasars, and cosmology, can attempt only to give an outline of the mass of research now in progress. Some idea of the intensity of effort now being devoted to radio astronomy may be gathered from the last two chapters, which are devoted to a description of radio telescopes and a list of the principal radio observatories of the world.

One sombre word of alarm must be added to any account of the achievements of radio astronomy. There are other uses of radio than the reception of information from outer space. The enormous growth of radio communications of all kinds has depended on and developed with the same technical advances that the radio astronomers have used and, in some cases, originated. The same receivers and aerial systems which have allowed radio astronomers to stretch our perception to the furthest limits of the universe are now used for radar, for navigation, and above all for television. There is an obvious clash of interests. On the one hand, the astronomers would like to see large parts of the radio spectrum kept clear of man-made transmissions, while on the other hand the communicators are finding great difficulty in accommodating their new systems within the whole of the available spectrum. The clash is becoming acute with the development of broadcast transmissions from satellites. There will soon be no place on earth out of sight of these transmitters, which are so 'radio-bright' that they can metaphorically turn night into day and spoil all further chances of radio astronomy. There is already a strict control on the use of radio frequencies, and some bands are allocated exclusively to radio astronomy. But if radio astronomy is to have a future as well as a history an even more determined vigilance will be necessary. Most urgently, the broadcast satellite transmissions must be kept well away from the radio astronomers' frequency bands; looking further ahead, there must be an agreed restraint in the use of transmitters throughout the electromagnetic spectrum, which should be treated as a natural resource to be carefully conserved. Perhaps this book will show that radio astronomers are making good use of that part of it which is still available to them.

CHAPTER 1

The Origins of Radio Astronomy

RADIO astronomy, a very young science, is a vigorous offshoot of a discipline whose roots are found in our earliest recorded history. Those who first observed and wondered at the courses and constellations of the stars, and the wanderings of the planets, had no instruments to extend their observation beyond the perception of keen eyes. But in 200 B.C., in Babylon, astronomers had already established a system of time-keeping based on their observations of the sun and the stars, dividing the year into twelve months, and the day into hours, minutes, and seconds. And sailors of the same period had learned to navigate with the help of the stars.

Even earlier, probably before 1000 B.C., a massive and highly organized effort had been put into the astonishing stone observatories which abound in Britain and Brittany. Stonehenge, Callanish, and the hundreds of lesser stone circles and alignments, are monuments to an age of man when astronomy must have been a dominating science and an unparalleled influence on human behaviour. The intention of these observatories was to follow the seasons of the year by keeping continuous watch on the rising and setting of the sun in relation to the bright stars. Possibly also some attempt was made to follow the more complex motion of the moon, and even to predict solar eclipses, although this is very uncertain.

In the face of Stonehenge it is humiliating to realize that an incomparably greater human effort has gone into the profitless field of astrology than into the science of astronomy. The elaborate observations and predictions of planetary and lunar motions which dominated astronomy for so long have often been used only to predict the character and fortunes of individuals born, conceived or married under a particular combination of heavenly circumstances. But the misguided observations and calculations of astrology did provide some of the material for a rational understanding of the dynamics of the solar system, which formed the basis of modern astronomy.

The practical importance of time-keeping and navigation ensured that positional astronomy was treated as a serious study for its own sake, especially in maritime countries. Even before the invention of the telescope, the accurate observations of Tycho Brahe of Denmark enabled Kepler to formulate precise laws of motion of the planets. On the basis of these observational laws, Newton was able to show that the force of gravitation was a universal force, applying as well between the moon and the earth, and the planets and the sun, as it does between the earth and an apple. With this new concept, the notion that man held a special position at the centre of the universe vanished, and the modern inquiry into the nature of the heavens began.

We may ask how it comes about that astrophysics is such a young subject when astronomy itself is 4 000 years old. But an inquiry into the nature rather than the arrangement of the stars requires tools that extend our perception of the universe beyond simple visual observations. The first great step towards knowledge of the stars themselves came with the invention of the telescope in the time of Galileo, although nothing was known even of the distance of any star until about 130 years ago, when the first parallax measurement of distance was made. To contemplate the composition even of the members of the solar system must at that time have seemed unthinkable. Dynamical studies gave eventually the masses and average densities of the planets, but nothing could be known of their chemical composition, or of their temperatures, or indeed of physical conditions on any heavenly body. The biggest revolution in astronomy was the introduction of spectroscopy, which came like a key to unlock all mysteries.

It was again Newton who first showed that white light can be split into many colours, all of which are simultaneously present in the white light, and who explained the action of the spectroscope which is used in every observatory for this purpose. In 1802 Wollaston showed that sunlight, when split up in a prism spectroscope, produced a spectrum in which there were narrow dark lines, the lines which are named after Fraunhofer, a later discoverer of the same phenomenon. G. G. Stokes was able to explain later that the light that should have filled these gaps in the sun's spectrum had been absorbed by atoms in the hot atmosphere surrounding the sun. Each separate element absorbed its own colour or wavelength, and the wavelength peculiar to each could be determined in the laboratory by measuring the spectrum of light emitted by electric discharges in vessels containing some of each element in turn. The Fraunhofer lines thus provide nothing less than a list of the kind of atoms to be found in the sun's atmosphere, and hence in the substance of the sun itself. Perhaps the most outstanding success of spectroscopy was the discovery by Lockyer, in 1878, of a strong Fraunhofer line in the green part of the solar spectrum at a wavelength corresponding to no known element. He suggested that it came nevertheless from an element, present in the sun but not at that time isolated on the earth, and he named it 'helium'. This gas is, of course, familiar today for its use in balloons, for many industrial uses, and as an end product of the fusion of hydrogen nuclei.

With the additional tool of spectroscopy, astronomy in the middle of the nineteenth century had acquired a new breadth in looking at the heavens. The view had changed from a peep through a narrow crack to a broad vista through the wide window of the visible spectrum. And extending to wavelengths longer and shorter than those of visible light, into the infra-red and ultraviolet wavelengths, the use of the spectroscope with photographic plates provided a source of information inaccessible to visual observation. In his famous treatise, published in 1873, Clerk Maxwell was able to show that these light waves were not the only form of electromagnetic radiation, but that there must be a wide range of wavelengths either side of the optical spectrum. Inspired by Maxwell's work, Heinrich Hertz made his famous discovery of radio waves in 1887. It was a long time before astronomers could detect and use these radiations, but with Hertz's discovery, radio astronomy became a possibility.

When radio was only three years old, the first recorded suggestion of the science of radio astronomy was made. It was a tremendous leap in the dark, and it needed a genius to take it. Thomas Edison was the genius and the suggestion was made in a letter to Professor Holden, Principal of Lick Observatory in California, written on 2 November 1890 by Professor A. E. Kennelly. Professor Kennelly, who first predicted the existence of the terrestrial ionosphere, worked with Edison in his private laboratory. The letter contains this sentence:

Along with the electromagnetic disturbances we receive from the sun which of course you know we recognize as light and heat (I must apologize for stating facts you are so conversant with), it is not unreasonable to suppose that there will be disturbances of much longer wavelength. If so, we might translate them into sound.

The means of translation was bold and ingenious, and is explained in the letter, quoted now at greater length.

I may mention that Mr Edison, who does not confine himself to any single line of thought or action, has lately decided on turning a mass of iron ore in New Jersey, that is mined commercially, to account in the direction of research in Solar physics. Our time is, of course, occupied at the Laboratory in practical work, but in this instance the experiment will be a purely scientific one. The ore is magnetite, and is magnetic not so much on its own account like a separate steel magnet but rather by induction under the earth's polarity. It is only isolated blocks of the ore that acquire permanent magnetism in any degree. Along with the electromagnetic disturbances we receive from the sun which, of course, you know we recognize as light and heat (I must apologize for stating facts you are so conversant with), it is not unreasonable to suppose that there will be disturbances of much longer wavelength. If so, we might translate them into sound. Mr Edison's plan is to erect on poles round the bulk of the ore, a cable of seven carefully insulated wires, whose final terminals will be brought to a telephone or other apparatus. It is then possible that violent disturbances in the sun's atmosphere might so disturb either the normal electromagnetic flow of energy we receive, or the normal distribution of magnetic force on this planet, as to bring about an appreciably great change in the flow of magnetic induction embraced by the cable loop, enhanced and magnified as this should be by the magnetic condensation and conductivity of the ore body, which must comprise millions of tons.

Of course, it is impossible to say whether his anticipation will be realized until the plan is tried as we hope it will be in a few weeks. It occurred to me that, supposing any results were obtained indicating solar influence, we should not be able to establish the fact unless we have positive evidence of coincident disturbances in the corona. Perhaps, if you would, you could tell us at what moments such disturbances took place. I must confess I do not know whether sun spot changes enable such disturbances to be precisely recorded, or whether you keep any apparatus at work that can record changes in the corona independently of the general illumination. I have no doubt however, that you could set us on the right track to determine the times of disturbance optically to compare with the indications of Mr Edison's receiver, assuming that it does record as we hope.

Of the experiment itself there appears to exist no record, except that, on 21 November, Kennelly writes that the poles have arrived ready to be set up. We may be sure, however, that no solar radiation was detected, for two good reasons. Firstly, very sensitive detectors are needed for even the strongest solar radio waves, and secondly, radiations of a wavelength long enough to be picked up by such an apparatus are prevented, by the terrestrial ionosphere, from reaching the earth. The experiment, though without hope of success, is admirable in its conception, and the idea that prompted it remains as the basis of a new science.

Although Edison was the first, he was certainly not the only famous scientist of the last century to attempt to pick up radio waves from the sun. In 1894 Sir Oliver Lodge said, in a lecture before the Royal Institution in London, that he was proposing 'to try for long-wave radiation from the sun, filtering out the ordinary well-known waves by a blackboard, or other sufficiently opaque substance'. He reports the experiment in a book called *Signalling across Space without Wires*:

I did not succeed in this, for a sensitive coherer in an outside shed unprotected by the thick walls of a substantial building cannot be kept quiet for long. I found its spot of light liable to frequent weak and occasionally violent excursions, and I could not trace any of these to the influence of the sun. There were evidently too many terrestrial sources of disturbance in a city like Liverpool to make the experiment feasible. I don't know that it might not possibly be successful in some isolated country place; but clearly the arrangement must be highly sensitive in order to succeed.

He has our sympathy. Optical astronomers suffer from city street lights; radio astronomers suffer from city street trams, and, in these days, from far worse evils in the form of radio-communication transmitters, whose frequencies range through the whole available radio spectrum.

Leaving aside these early attempts, we can assign a birthday to our subject with an accuracy unusual in science. In the issue of the *Proceedings of the Institute of Radio Engineers* of December 1932, Karl Jansky published an account of his historic first reception of radio waves from outer space. Had 1932 been a year of high sunspot activity, Jansky would undoubtedly have found the radiations from the sun for which Edison and Lodge had looked in vain. As it was, the sun was quiet, and instead the radio waves coming from our galaxy were discovered.

Jansky was carrying out for the Bell laboratories in America a study of the noise level to be expected when a sensitive short-wave radio receiver is used with a directional aerial system in long-distance communications. He was listening for crackling noises from thunderstorms, and he had an aerial system which could scan round the sky to find the direction from which these signals came. It was rather a clumsy contraption, mounted on Ford model-T motor wheels for easy rotation. The noise level was found never to decrease below a certain level, but this level varied gradually through the day. When Jansky listened on an ordinary loudspeaker to the output from his receiver, he heard only a steady hissing sound, quite different in character from the crackles of thunderstorms.

Jansky found that the greatest signal always occurred when the aerial pointed in a certain direction in space, that is in a direction fixed relative to the stars, and not relative to the earth or even to the sun. This direction turned out to be the direction of the centre of our galaxy, where there is the greatest concentration of stars, and Jansky was able to say, without any doubt, that he was listening to signals 'broadcast' from the Milky Way.

Jansky's discovery was well publicized in America. Alongside the technical reports there were newspaper reports, and even a radio programme broadcasting to the people of America the recorded hissing sound that came from Jansky's receiver. Surprisingly enough, however, the subject was dropped almost completely after Jansky had carried through the original plan of his experiment, and it was not until after the Second World War that radio astronomy was established as a separate science, and started its spectacular growth.

One radio amateur bridged the gap between Jansky's first discovery of radio signals received from outer space, and the post-war surge into the new science. This was Grote Reber, who for several years was the only radio astronomer in the world. He is still a radio astronomer, and he is still a pioneer, exploring new branches of the subject with the same enthusiasm which he once applied to amateur short-wave radio, and then to the construction of the first radio telescope.

Reber accepted Karl Jansky's work as an opportunity and a challenge. He built for himself, in his spare time, with his own resources and in his own back yard in Wheaton, Illinois, a steerable parabolic reflector, 30 ft in diameter. This he used, with sensitive receivers, to make some remarkable recordings of cosmic radio waves.

After many disappointments, he found that he could record the 'cosmic static', discovered by Jansky at a wavelength of 15 m, on the much shorter wavelength of 60 cm, where his new radio telescope was able to give precise details of the direction of origin of the radio waves. In his maps, which in many details remained unsurpassed for fifteen years, it is possible to see for the first time startling differences between the visible sky and the radio sky. Nothing could be seen of radio waves from any visible object, except the sun, but peaks of emission were to be seen in various parts of the Milky Way where there was no reason to expect them. Some of these peaks we can now recognize as the discrete radio sources in the constellations of Cassiopeia, Cygnus, and Taurus. Reber recognized the importance of his experiments, and in papers published in 1940 he showed quite clearly that a new branch of astronomy was in being.

Reber's results were published in 1940 and 1942, when the attention of most scientists was directed elsewhere than towards pure research. Nevertheless, his papers were not ignored, and towards the end of the war his results were gathered together with several interesting new observations made as a by-product of research in radar, and used as a stepping stone towards the beginning of radio astronomy as we now know it.

The contribution of radar research came largely through the work of one man, J. S. Hey. His wartime job was in the Army Operational Research Group, where with his colleagues he studied reports of the efficiency of all the army radar equipments, and investigated all the reports of jamming by enemy transmitters. From these reports came the first indication that the sun could transmit radio waves in the metre-wavelength region. This radiation, as we shall see, was radiation from an active sunspot, and it was the following up of this clue by new radio observatories at Sydney and at Cambridge that started the extensive work on the radio emission from the sun, both from spots and from the surface of the 'quiet' sun.

The second result from wartime radar experience was the detection of radar echoes from meteor trails. Hey and his colleagues were able to continue work with army radar sets immediately after the war, and to follow up reports of short-lived echoes whose association with meteors was only conjectural. Hey showed that every meteor leaves a trail of ionization in the upper atmosphere, which reflects radio waves as effectively as a long metallic wire. The operational necessity of tracking down the source of unwanted radar echoes was soon replaced by the driving curiosity of scientific investigation, for Hey found that the echoes not only could be used as a means of exploring the upper atmosphere but that they also revealed the existence of concentrated streams of meteors of greater intensity than any previously seen.

Hey's third contribution came as the result of an experiment planned from the outset as a work of observational radio astronomy. Following Jansky and Reber, he decided to plot out the cosmic radio waves over the sky with the best aerial system available. He soon had a map of the radio emission from the northern sky, showing the Milky Way as its most prominent feature. But in one spot in the constellation of Cygnus he found a fluctuating signal, which he interpreted correctly as a radio signal from a discrete source. As was shown later, the fluctuations were imposed on the signal by the terrestrial ionosphere, while the source itself was a steady and powerful transmitter. It was some years before it was identified with the exciting object known as Cygnus A, now known to be one of the most powerful radio galaxies.

After the publication of Hey's work the progress of radio astronomy was very rapid. Nevertheless, it was scarcely recognized as an integral part of astronomy for about another ten years. By that time the optical astronomers began to realize that the cosmic radio waves originated in some of the most interesting objects in the universe, and that it might provide a powerful means for exploring the depths of space, possibly even for answering some of the most pressing problems of cosmology.

During this period the main development of radio astronomy was in Sydney, Australia, under J. L. Pawsey, and in Cambridge, England, under M. Ryle. Significantly, both groups were made up of physicists, not trained as astronomers but with the same wartime experience in radar that led Hey to his discoveries. Only one optical astronomer, J. H. Oort, provided any inspiration for the new subject at this time. The story of radio astronomy in Holland revolves around Oort even today, but his imaginative step of suggesting a search for a radio spectral line (Chapter 8) marks the beginning of a slow synthesis of the new and old techniques.

Today there are new techniques to add again, with the developments of infra-red, X-ray, and gamma-ray astronomy. Acceptance of these has been immediate, and it is now entirely natural to bring together information from the whole wide electromagnetic spectrum in trying to solve a problem in astrophysics. Radio waves, however, still have a most important part to play in providing observational information, because they originate in such different ways and different places from the light waves of optical astronomy. There is almost a different universe to be explored with our new radio eyes, filled with unfamiliar new objects, and with familiar objects strangely altered, revealing unexpected and often startling new characteristics.

CHAPTER 2

Cosmic Radio Waves

WHEN we look up at the sky on a clear night we look into a vast mystery. Our eyes receive as points of light the radiations from a myriad of stars, and as wonder turns to inquiry we realize that these radiations contain all the information we can ever receive from the visible universe. With reservations perhaps inside our own solar system, we cannot touch any part of it, we cannot perform experiments on it; we can only look at it.

The radio astronomer looking out into the universe receives the electromagnetic radiations known as radio waves instead of light. Although he uses complex instruments for analysing these radio waves, he still adopts the language of the visual observer, and talks of what he 'sees' with his radio telescope. He can build up a picture of the universe, a radio universe, just as real and meaningful as the familiar visible universe. Where the two pictures differ, they in fact complement one another. Together they provide our only knowledge of the universe around us, and we must at the outset recognize the limits of our perception. The limits are so narrow that we may doubt the reality of the fantastic universe so imperfectly revealed.

Dr Johnson, when he was confronted with the argument that a stone could not be said positively to exist except as an impression in his brain, so that all that he could logically say about the stone was that he *thought* he saw it, said 'I refute it thus', and kicked the stone. No refutation really, but only a demonstration that all our knowledge of nature comes via our senses, normally more than one sense. For proof of the existence of the stone, the sense of touch is almost as good as the sense of sight. Our total detailed information on the nature of the stone, and of any other physical object, depends very much on the different senses we use in our inquiry about it, and on the way in which those senses are used. Putting it in a different way, physics is the acquisition and marshalling of information about the universe, and in all parts of

24

physics the first emphasis must be on the nature of experimental observations and their interpretation. What is the nature of the information available to astronomy?

The whole science of visual astronomy has grown from optical observations of light radiated from heavenly bodies, and this light brings information in several ways. Firstly the bodies can be hot, and shine with their own light. The quality and quantity of this radiation tells us much of the composition and the temperature of its source. Secondly, light may be absorbed, perhaps totally, as when Venus or Mercury is observed in transit across the disc of the sun, or perhaps only partially, indicating the existence of a cloud of gas or dust in space. When light is partially absorbed, it is often found that its quality, or colour, is changed, and also that its spectrum shows absorption lines at particular wavelengths corresponding to the atomic species present in the cloud. The extent of the absorption, and the wavelengths of the absorption lines, tell us the density and composition of the absorbing medium. Thirdly, light can also be reflected, either specularly or diffusely; by reflection of sunlight we are able to see the planets and the moon, and by the reflection of starlight we are able to see some of the clouds, or 'nebulae', in our galaxy which have not enough energy to shine with their own light.

Lastly, light can suffer refraction, between its source in the heavens, and our eyes or optical instruments. This 'bending' of the light waves takes place in the earth's atmosphere and may be a nuisance in optical astronomy, but in studying the terrestrial atmosphere, the random refraction and diffraction effects which make the stars twinkle can be turned to good account.

As well as the light waves from the sun, moon, stars, and planets, there are two kinds of visitors to this planet from outer space, bringing their own rather more direct information. These are the cosmic rays, individual atomic and sub-atomic particles continually bombarding the earth, and originating partly in the sun, partly in the galaxy, and probably partly even from beyond our galaxy; and the meteors and comets, objects from many tons weight down to tiny dust particles, most and probably all of which are part of the solar system. Of these, the cosmic rays have a special interest for the radio astronomer, as they are the high-energy particles responsible for the radiation of radio waves in many of the most interesting objects in the radio universe.

Knowledge gained by radio astronomy must come via channels similar to those of optical astronomy, i.e. radiation, absorption, reflection and refraction. The essential difference in the kind of information received is that the regions and objects which radiate, absorb, reflect, or refract radio waves, appear often to be very different from the objects of the visible universe. Using radio waves, we can 'see' such invisible objects as a cloud of electrons, or of transparent hydrogen gas.

The Nature of Radio Waves

Sound and television broadcasting have made communication by radio waves a household commonplace, but it is still difficult to grasp the idea of energy propagated through space, in the form of oscillating electric and magnetic fields, without the support of any medium such as air or an ionized gas. Broadcasting has, however, made us familiar with two important concepts, wavelength and polarization; the wavelengths of the various television bands are obviously related to the length of the dipole rods used in receiving aerials, or 'antennae', while the different orientations of the antennae for 'vertical' and for 'horizontal' polarization correspond to two possible orientations of the electric field in the electromagnetic wave. Television broadcasts use wavelengths between 6 m and 30 cm. Wavelength is related directly to the frequency of the oscillations in the wave; television uses frequencies from 45 MHz to 800 MHz. (The unit MHz stands for Megahertz, or one million oscillations per second.)

Cosmic radio waves can be detected over a very wide wavelength range, which is only limited by the absorption of the atmosphere at wavelengths shorter than a few millimetres, and by reflection from the ionosphere at wavelengths longer than about 30 m. Most cosmic radio waves are not polarized, and their energy can be detected by antennae whose dipoles are at any orientation. There are, however, some particularly interesting cosmic radio sources which emit linearly polarized radiation, analogous to the polarized radio waves used in television. Another possible state of polarization, known as circular polarization, is also of some importance; in this state of polarization the electric field does not remain in one fixed plane but rotates uniformly at the frequency of oscillation.

The specification of a radio signal involves the wavelength (or frequency), the polarization and the strength. The strength is

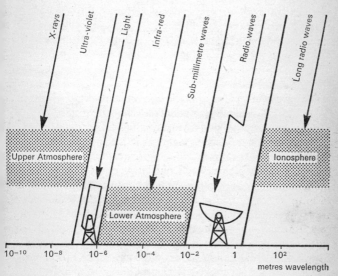

1. The spectrum of electromagnetic waves. Only two bands of wavelengths, light and radio, can penetrate the terrestrial atmosphere and ionosphere

measured as a flow of energy across an area, as for instance the flow of energy into the large collecting area of a parabolic radio radio telescope. The energy is spread over a range of wavelengths, and therefore also spread in frequencies, so that the 'bandwidth' of the energy flow must be specified. The unit needed is therefore one of 'flux density'; this is measured in watts per square metre (watts m^{-2}) per unit of frequency bandwidth. In practice the strength is so small that flux densities are commonly quoted in units of 10^{-26} watts m^{-2} per unit bandwidth.

The way in which radio emission from a source is spread over a

27

range of frequencies is described as a spectrum, analogous to the spectrum of visible light. Most sources radiate a spectrum which is continuous over the radio range, but some radiate predominantly in narrow frequency ranges. These are the frequencies of the radio spectral lines, such as the line of 1 420 MHz (wavelength 21 cm) which is characteristic of neutral hydrogen.

The Sky Seen through Radio Eyes

Looking out now at the radio universe beyond the terrestrial atmosphere, we may imagine that our eyes are no longer sensitive to light, and that we can see radio waves instead. Let us also imagine that it is daytime, and that the sky is overcast so that the optical astronomers are able to relax, perhaps disappointed that the chances of observing are slender for the night to come. We look up and find that the clouds are transparent and the sky is as clear as on the brightest of starlit nights. And it is indeed starlit, as the sun has become a strange dim object, with the blurred outlines of the corona showing clearly around it. The moon and the planets may be faintly visible, but quite inconspicuous. The familiar bright stars cannot be seen at all, except for a few dim flickers from some of the nearest.

The outstanding feature of the radio sky is the Milky Way, seen in its accustomed place in the sky but shining splendidly and prominently, brightest in the southern sky where it is seen close to the centre of our galaxy, but extending as a bright band right across the sky. Like the solar corona, its outline is blurred. It is made up not of many separate stars but of glowing hot gas filling most of the space between the stars. In it we see some bright star-like objects, which on closer inspection are seen to be clouds of gas rather than solid stars. Many of these are nebulae familiar to optical observers, but with strangely different shapes and brightnesses. A line of them shines like a row of pearls along the centre line of the Milky Way.

We look in vain for the familiar stars of the constellations such as Orion, Taurus, the Plough. Instead we see quite different star-like objects covering the whole sky, outside the Milky Way as well as in it. These again are not stars, but other more distant galaxies like our own, more prominent to radio eyes than optically. These are the radio galaxies and the quasars, in which processes unknown in our own galaxy are feeding vast reserves of electrical and magnetic energy into the radio part of the spectrum.

The final surprise is that the radio sky is changing. Radio waves pick out the variable objects; these are often exploding violently, sometimes as individual stars but more usually as the central cores of galaxies. This violence manifests itself in radio emission, since radio comes naturally from the hot ionized gas surrounding exploding objects; furthermore this gas, and the radio waves it emits, can change much faster than the more condensed objects which radiate most of the starlight. The radio sun is extremely variable: at times it can burst out with dazzling brightness on the longer radio wavelengths. The planet Jupiter spasmodically sends out bursts of radio impulses, as though a violent thunderstorm were raging on its surface. Even the most distant extragalactic objects, the radio galaxies and quasars, may have large variations; these are usually observed mainly at the shorter radio wavelengths.

The most remarkable of the variable radio sources appear as regular flashes of radio signals, like lighthouse beacons. These are the pulsars, which are now known to be associated with a previously unobservable type of star, the old and incredibly condensed neutron stars. The pulsars are distributed throughout our own galaxy, so that they are seen concentrated in the Milky Way. Their flashes occur at various rates, from 30 per second to 1 every 4 seconds. If ever we need a navigation system for a space cruise among the stars of the Milky Way, the pulsars will provide the navigational beacons.

Thermal and Non-Thermal Radiation

Why does the sky look so different for 'radio eyes'? It must first be realized that most of the difference lies not in an unnatural dimness of the familiar objects, but in an unexpected brightness of unfamiliar ones. Light is radiated by most of the heavenly bodies because they are hot, and the amount of light they radiate is determined by their temperature. This is thermal radiation, which extends throughout the electromagnetic spectrum but with an intensity that depends on wavelength. (It is generally more convenient to specify the spectrum in terms of frequency than wavelengths, remembering that the product of wavelength and frequency is a constant, which is the velocity of light.) The ratio between radio- and light-emission from a hot body is given by the Rayleigh–Jeans Law:

The energy per unit frequency range is proportional to the square of the frequency.

Now the frequency of radio waves is about ten million times less than that of light waves, so that this law tells us to expect very little indeed in the way of radiation from the sun, or any other hot body, at radio wavelengths.

Little as it is, this thermal radiation can be detected from the surface of the sun, from the planets, and from some clouds in the Milky Way. Short radio wavelengths, i.e. high frequencies, are best for this since at the lower frequencies the weak thermal energy is entirely swamped by very much stronger radiation processes.

When radiation can be recognized as thermal, its strength can be used as a very direct measure of the temperature of the emitter. For example, the strength of the radio emission from the moon at high radio frequencies tells us that the average surface temperature is about 190 K, while the temperature varies by about 30 K above and below this value through the lunar month. (These temperatures are measured on the 'absolute' scale, on which zero is the lowest temperature theoretically possible. A temperature of 190° absolute, or 190 K, is about the lowest temperature one would ever expect to find on earth. It corresponds to $-83°C$ or approximately $-118°F$.)

Again, the temperature of the neutral hydrogen gas in interstellar space can be measured from the strength of its characteristic radiation at 1 420 MHz; this is thermal radiation, even though it only occurs at the frequency of the characteristic spectral line. Most of the hydrogen is at temperatures between 40 K and 120 K.

Much higher temperatures occur close to stars. The surface of the sun is at a temperature of 6 000 K, while the solar corona is at a temperature of about one million degrees absolute. Both these temperatures can be measured from the strength of thermal radiation, respectively from the surface and from the much more extensive corona; distinguishing between the two sources is, however, rather complicated. The problem is discussed in Chapter 4.

There is usually no need to worry about the precise mechanism of radiation, provided that one can be sure that it is indeed thermal. The essential point about thermal radiation is the precise correspondence between its strength and the temperature of the emitter. In the examples of thermal radiation quoted above the origins are respectively in the solid surface of the moon, in the line radiation of neutral hydrogen gas, and in the broad-band radiation of ionized gas in the solar corona. The detailed processes differ, but the result is the same.

Non-thermal radiation is responsible for the whole brilliance of the radio sky, providing all of its unfamiliar features. The Milky Way, which is the disc of our galaxy seen from near one edge, produces thermal radiation from neutral hydrogen, and from some clouds of ionized hydrogen which resemble the solar corona; the hydrogen atoms and electrons which emit this thermal radiation have energies which are related to the temperature of their own regions of the Milky Way. But pervading the whole of interstellar space is a chaotic stream of particles with very much higher energies, quite unrelated to the temperature. These are the cosmic rays, observed on earth by their passage through photographic plates, or cloud chambers, and other detecting devices familiar to atomic physics. In a sense their radio radiation is thermal, as its energy derives from the individual motions of these electrons which have a kind of temperature of their own, but there is here no real thermal equilibrium between the radiation and any body of hot gas. The distinction between the energies of the cosmic rays and of the cold and sluggish gas through which they pass is obvious.

In the galactic plane there are some especially active spots forming discrete sources of radio waves, and these again are non-thermal radiators. The energy supply for these is fairly evident, as many of them are visible nebulae which are the remains of exploding stars or 'supernovae'. The visible wisps of gas produced by these supernova explosions are still blowing outwards at some hundreds of kilometres per second. One such nebula, the Crab nebula in Taurus, contains cosmic-ray electrons which are such powerful generators of non-thermal radiation that they produce visible light in addition to radio waves.

Looking further out into space, the radio sky contains many bright nebulae, each of which is a complete galaxy of stars like our own. Some, like our own galaxy, radiate strongly by virtue of the electrons circulating throughout the interstellar space; others, which may be seen at very much greater distances because of their greater power, appear to be in a state of catastrophic explosion. These are the radio galaxies and quasars which are the most exciting features of the whole radio sky, for it seems that they may be so powerful that their non-thermal radiation can be 'seen' even beyond the distance covered by the 200-in telescope of Mt Palomar, and they may be the only means of revealing to us the structure of the most distant parts of the universe.

CHAPTER 3

The Radio Universe

LET us now take a preliminary reconnaissance through the universe which is opened up by our new 'radio eyes', starting with the solar system, then proceeding outwards through our galaxy, the Milky Way, to the extragalactic radio sources and the cosmological information they contain.

THE SOLAR SYSTEM

The Sun as a Radio Transmitter

The sun must be accounted to be the most important of the heavenly bodies which illuminate the radio sky. The radio sun is not so prominent as the visible sun, nor indeed do its radio waves directly affect our daily lives except as indicators of solar disturbances, which in turn may affect radio communications or compass bearings on earth. Nevertheless it is an extraordinarily rewarding object to study by radio, with a wealth of detail on the surface at any one time, and with a very variable behaviour. The sun is not an uncommon type of star; there are many millions of stars like the sun in the galaxy, but the next nearest one to the earth is so much farther away that its radio waves are undetectable.

The sun represents a laboratory where conditions quite unrealizable on earth are accessible to our radio telescopes. Different parts of the sun have different conditions; we can at any time look at radiation from the surface at a temperature only existing on earth in the fireball of a nuclear explosion, or we can observe the effect of electrons spiralling freely round lines of magnetic force in a gas more rarified than the best laboratory vacuum. The radio waves these electrons emit are often the only clue to their very existence, and the only indication of the forces which accelerate them to energies sometimes as high as those of cosmic rays. We shall see how it is that by measuring the radio waves emitted by the sun and the corona around it we are able to explore a wide range of physical conditions, and to connect some of the spectacular events on the sun with geophysical phenomena such as aurorae and magnetic storms.

The radio transmitters which broadcast the BBC programmes do so by generating oscillating electric currents in aerial wires held high above the ground, so that the radio waves are launched efficiently into space. The electric currents themselves consist of moving clouds of electrons in wires, or in electric discharges. It is the art of the radio engineer to regiment these movements in ways which do not produce heat, like current in an electric fire, but which produce radio waves from cold aerial wires.

In the solar atmosphere, the temperature is so high that the hydrogen gas of which it is largely composed is dissociated into protons and electrons, which are then kept in rapid motion by virtue of this same high temperature. It is for this reason not surprising that the sun is a radio transmitter, and indeed that it is the most outstanding radio transmitter in the sky. No self-respecting radio engineer would, however, encourage such physical conditions, in which the electron motions are entirely disorganized, moving in all directions in complete chaos. The difference from the orderly oscillation in a transmitting aerial is reflected in the difference in the radiation; oscillation at one chosen frequency produces only that one frequency of radiation, while random motion produces at once every frequency in the radio spectrum, with the energy of the electron gas spread widely instead of concentrated in frequency as it must be in a radio broadcasting station.

There are two main kinds of solar radiation. The sun is said to be 'quiet', when there are few sunspots and no flares to be seen. More usually it is 'active', with a very wide repertoire of behaviour. The quiet sun is a complicated enough radio transmitter, but, as we shall see in later chapters, the active sun outdoes it in every way – in spectacle, in complication, and in defiance of theory. The quiet sun is reasonably well understood, and the story told by the radio waves it transmits agrees with the story told by the light waves, and supplements it in a most satisfactory way.

The Planets

If we could see the solar system from outside, perhaps from a distance of a few light-days, the planets would optically be very inconspicuous in comparison with the sun, and the earth would be a very faint star-like object. The same would be true for our radio eyes, except that the earth itself would be recorded in the recent astronomical literature as a most exciting and explosively active nova. This would be the result of the man-made radio and television transmissions which now pour out at frequencies all through the spectrum used for radio astronomy. It is a sombre thought to consider the interpretations that might be put on these radio signals by the radio astronomers of another planetary system.

One other planet, however, is known to be active, although not through any intelligent agency. Jupiter is most surprisingly a source of intense bursts of radio emission, in some way like the radio 'atmospherics' produced by lightning flashes. The explanation is not at all clear, except that this intense radiation is evidently 'non-thermal', and that it originates in some organized motion of electrons in the Jovian magnetic field. Radio observations of Jupiter have in fact been able to tell us a great deal about this magnetic field, and about the rotation of the planet.

The planetary system has recently become the subject of rather more direct radio exploration than the simple reception of its naturally generated signals. Radar can now be used to measure the distances of several of the planets with a remarkable accuracy, completely out-classing the previous measurements of the size of the system and the detailed orbits of the individual planets. We shall include some of these results in our later more detailed discussions. Even more direct exploration, by spacecraft carrying television cameras and other scientific instruments, has now reached the planets Venus and Mars; within a short while Jupiter will be reached, and there is at least a technical possibility that most of the outer planets can also be visited by one of these 'deep space' probes.

Space research, in this sense, is closely related to radio astronomy in several ways. Most obviously, the radio signals transmitted by a distant spacecraft are so feeble by the time that they reach us that they can only be received by a large and sensitive radio telescope. Many radio telescopes in fact serve a double duty; either the spare time of a space telemetry station is used for radio astronomy, or a large radio telescope devoted to astronomy is borrowed for a few days when the most critical part of a space mission demands the utmost sensitivity. There are, however, some basic scientific investigations which involve only the transmitted radio signal, as contrasted with the information it carries.

Asteroids, Comets and Meteors

The planets may be regarded as the big fish of the solar system; there are also several varieties of small fry to consider. There is first a gap in the sequence of planetary orbits, between Mars and Jupiter, where instead of a planet there is a swarm of tiny planet-like bodies called the asteroids. The largest, Ceres, is 8 000 times less massive than the earth; the total mass of the tens of thousands that can be seen amounts to less than 1 per cent of the earth's mass. Asteroids are small, cold, and have no atmosphere. They therefore make no impact on our new radio eyes, and we pass them by. Comets again are not suitable subjects for radio astronomy; this is less to be expected since they notably possess a long glowing tail which looks more like the hot gas which we expect to radiate radio waves. The tail of a comet turns out, however, to be cold and very tenuous, and it cannot even be detected by its effect on radio waves passing through it from more distant sources.

The smallest solid bodies in the solar system are those that reach the atmosphere and burn with a bright trail as 'shooting stars' or meteors. These bodies mostly weigh less than a gram, but their velocities are so great that they have a tremendous energy to impart to the air. The result is not only a brilliant streak of light, but a dense trail of ionized gas high in the atmosphere. This trail can be detected by radar.

Meteor astronomy is perhaps a side branch of modern radio astronomy, but its central importance in the early history of radio astronomy ensures a place for it in this book. The minor bodies of the solar system will therefore all be included with the meteors in Chapter 15.

The Solar Corona and Interplanetary Space

The way in which our 'radio eyes' pick out a thin hot gas in preference to solid bodies is nowhere more marked than in the solar system, where the familiar planets, asteroids and comets are either inconspicuous or undetectable while by contrast the apparently empty space surrounding them proves to be a very lively place. The solar corona can be seen optically to extend at least to several times the diameter of the sun; the quiet radio sun, as a transmitter, extends to about the same distance. The active sun, however, stimulates low-frequency radio emission from local regions of the corona at much greater distances. These bursts extend to very low frequencies, in the region of 1 MHz, where they can only be detected from satellite-borne receivers, since they do not penetrate our terrestrial ionosphere.

A far more potent investigation can be made of the corona, extending at least to the distance of the earth's orbit, where it becomes the 'interplanetary medium', by observing its effect on the transmission of radio waves through it. Again the main effects are seen at comparatively low radio frequencies, but there is no need to use such low frequencies that satellite-borne receivers are necessary.

There are plenty of radio transmitters available for this work: they are the 'radio stars' of the sky, the discrete sources which we shall be describing later in this chapter. When the radio waves from these sources pass through the interplanctary gas, they are scattered and diffracted so that they spread in angle, possibly only by a few seconds of arc. A radio source, whether it is a quasar or a pulsar, or any other effectively 'point' source, then appears as a slightly blurred disc. Furthermore, the disc is not static: it seems to be dancing about and changing size, just like the twinkling of a visible star. This is the effect of small moving irregularities in the gas. The speed and direction of these movements can be measured; they are found to be part of a continual outward flow from the sun, now known as the solar wind.

We shall encounter this phenomenon of radio twinkling, or 'scintillation', again in the interstellar gas of the Milky Way (Chapter 8). The solar wind also has a wide importance, since it has a controlling influence over magnetic conditions in the region of the earth. Its strength may be judged from the appearance of the tails of comets, which always point away from the sun (Plate XVI). These tails are acting like weathervanes, forced to point always down stream in the powerful wind blowing out through the lonely regions of interplanetary space.

THE MILKY WAY

When we turn our thoughts from the sun to the powerful radio waves which Jansky found to be radiated by the Milky Way, we are considering a region where distances pass outside the range of everyday understanding. Planets move round the solar system in times of less than a human life-span, and we can even fire rockets from the earth to form artificial planets. But the nearest stars outside the solar system are so far from us that even light takes several years to make the journey from there to the solar system, and light travels at 186 000 miles per second, in contrast to the speed of ten miles a second which may be reached by rockets.

The nearest stars show no particularly orderly grouping in the sky, but the further ones which are very much fainter and very much more numerous combine together to form the familiar bright streak of the Milky Way. This tight group of stars is surrounded at greater distances again by a vast emptiness, in which other blobs of light are to be seen, each containing thousands of millions of stars in a compact bunch. Each is called a galaxy. Our present understanding of the size and distance of these galaxies is of quite recent origin. Radio astronomy may prove very useful in studying the most distant of them, as we shall see later, but it has proved its worth above all in the study of our own galaxy. The great penetration of radio waves allows the radio astronomer to explore the whole depth of the Milky Way, where most of the stars of the galaxy are concentrated, and where a concentration of dust obscures light so severely that in some directions all but the nearest stars are blotted out. Furthermore, radio waves have been found to come from a great halo of space around the galaxy, where no stars exist at all.

In Jansky's accounts of his early work on radio waves from the galaxy occur two rather speculative remarks on the origin of the radiation. Firstly, he suggested that the distribution of brightness over the sky agreed rather well with the familiar appearance of the Milky Way, so that the radio sources might be distributed in space in a disc, like the disc which contains most of the visible stars in the galaxy. Secondly, from the character of the signals, which sound just like the 'noise-like' radio waves produced by thermal radiation, he suggested that the origin might lie not in the visible stars, but in the tenuous ionized gas lying between the stars, a gas that might be kept hot by radiation from the stars. We shall see that Jansky's speculations fit modern observations remarkably well, although some surprising new features have come to light in the radio sky.

Before describing the appearance of our galaxy as we see it through a radio telescope, it is as well to recall the picture given by optical studies. The stars which are outstandingly bright to the naked eye are only tens or hundreds of light-years away, which is near to the solar system in terms of the scale of distance we must use before the structure of the galaxy becomes apparent. And therein lies a great difficulty. The galaxy also contains interstellar material in the form of dust, which obscures the view in the very directions in which we want to see the farther reaches of the galactic structure. In fact, the story of galactic structure must begin, not in our own galaxy which is partially hidden from us, but in other more clearly visible galaxies, the spiral nebulae, at distances from us a hundred times as great as the distance between our earth and the centre of our own galaxy.

Pictures of extragalactic nebulae, such as those in Plates I and II, are familiar to us today; it is surprising to find, however, that at the beginning of this century, even the rough shapes of these nebulae were hardly known. There was indeed only a suspicion that their distances were greater than those of most other visible objects. It was the work of Edwin Hubble (1889–1953) at Mt Wilson Observatory which established that these nebulae were composed of stars like the stars of our own Milky Way, and it was Hubble who first understood clearly that these were complete entities, floating isolated from one another by vast depths of empty space. He named them 'island universes'.

There are many different kinds of extragalactic nebulae, and not all of them are spiral nebulae, but it happens that the nearest spiral to us is one which also bears a very close resemblance to our galaxy. This is the Andromeda nebula (Plate I). Comparison with other nebulae shows that the oval shape is an effect of perspective, and that the nebula is shaped like a disc, in which lines radiate roughly spirally from a concentrated centre, tracing the regions where bright stars are to be found along with clouds of gas and dust.

In our galaxy, the sun is embedded in a similar disc of stars and dust, and it lies in one of the spiral arms about two thirds of the distance from the centre. The bright stars we see around us are mostly in this arm, and consequently are to be seen in every direction. Only in the direction of the galactic plane are there very many distant stars to be seen, and here also the effects of dust obscuration can be seen as dark rifts in the bright background of the Milky Way. In the Andromeda nebula, the dust is not so effective a screen, as it can at most cut off the light from the further half of each spiral arm in the nebula, leaving us a complete picture of one side.

Because the solar system is embedded in interstellar dust, hiding most of the distant stars, our own galaxy has to be examined by less direct methods. A classical problem in dynamical astronomy has been to use the observed motion of the nearby stars, relative to the sun, to derive the distribution of mass in the galaxy: the stars are moving round the disc in circular orbits, at speeds which depend on the mass inside their orbits. The velocities of stars are deduced, partly from their apparent motion across the skies, and partly from their radial velocities as measured by the Doppler shift of spectral lines. It has proved quite easy to find from these velocities the total amount of matter in the galaxy, but it is on the other hand rather hard to provide more than a general law of the variation of density with radial distance from the galactic centre, even taking no account of the variations in thickness. This cannot take us very far in the examination of our own galaxy, since at great distances many stars in the vital directions in the galactic plane are obscured, and the movements of those that are visible are small and hard to measure. Nevertheless, a distribution of stars in the disc can be found by this method, and can be shown to resemble that of the Andromeda nebula.

Again, the distances of stars can be measured by observing the intensity of their light, their optical 'magnitude'. Provided that allowance can be made for obscuration in the dust clouds, this depends only on the type of star and its distance from us. This technique applied to some particularly bright stars – the O and B stars found in the spiral arms of the extragalactic nebulae – provides some idea of the distribution of these kinds of star in the disc. A definite indication of lanes of stars running tangentially to the radius of the galaxy can be found, just like the arms of a spiral galaxy.

The plane of a galaxy is not, however, the only place where stars are to be found. Some galaxies, indeed, have little or no disc, and appear to consist of a spherical or elliptical blob of stars. An example can be seen in Plate III. Our galaxy contains some types of stars which are found in a similarly widespread distribution, concentrated in the centre, but extending out in all directions from it. Clusters of stars, known as the 'globular clusters', are found at great distances from the galactic plane, and these do not contain the bright hot stars to be found in the spiral arms. The widespread halo, or corona, of stars, appears to be quite distinct from the plane both in shape and in population.

Astronomers have long classified stars into types, designated by initials, forming the sequence, O, B, A, F, G, K, M, R, N, S (traditional mnemonic: Oh! Be a fine girl, kiss me right now. Smack!), in which temperature decreases progressively from O to S. The sun, for example, is a G-type star, neither remarkably hot nor cold. The O and B stars are those which are hot enough to heat the interstellar gas all round them, and produce the H II regions (Chapter 8), which are found in the spiral arms. Other members of this main sequence of star types are also found in spiral arms, but they are rather less concentrated to the plane. A very different population of stars is to be found in the widespread halo of the galaxy, and this clear-cut distinction is now recognized in the designation Population Type I for the disc, and Population Type II for the halo.

The optical appearance of our galaxy from outside must then be somewhat as suggested in Figure 2, which shows a rough spiral structure, with the sun *S* in one arm, and a Type-II population of stars distributed thinly throughout a roughly spherical halo,

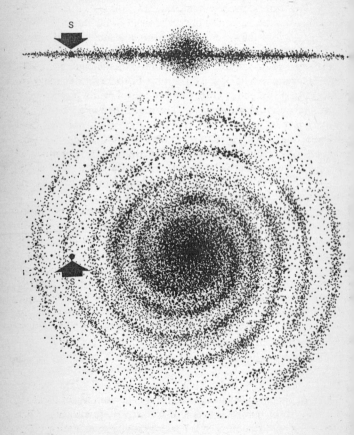

2. Sketch of a spiral galaxy (a) from the plane of the galaxy, (b) from above

concentrated on the centre of the galaxy. About one hundred thousand million stars make up the whole galaxy, with most of them lying fairly close to the galactic plane.

Radio Stars

When Hey first showed that the bright radio background of the Milky Way contained an individual, discrete source of radio waves in the constellation of Cygnus, and when other such discrete sources were subsequently found by the new radio observatories in Australia and in Cambridge, it was natural to think of them as radio stars. None of these early discoveries turned out eventually to be star-like; in fact most of them were extragalactic nebulae. Recently, however, a number of genuine cases of 'radio stars' have come to light (Chapter 6), some of them through bright flares of radio emission lasting some minutes or hours, others through discoveries of strong X-ray emission from some very inconspicuous stars which subsequently have been found to be weak radio sources. These occasional 'radio stars' are distinguished from ordinary stars by some process or processes in their atmospheres rather than in their solid surfaces.

Despite the generally poor showing of visible stars to our radio eyes, there is now a new class of stars, discovered through radio astronomy, and represented (in 1974) by over 100 examples of known radio emitters. Only one of these can be seen optically. They are the pulsars, which are now understood to be neutron stars, so small and cold that no detectable radiation of any sort could be expected from their surfaces. Here is one of the major discoveries of radio astronomy, which has revealed an entirely new physical situation in the atmosphere of these highly condensed, magnetized, and rapidly rotating bodies (Chapter 7).

Broad-Band Milky Way Radiation

Although the main feature of the radio sky is the bright band of the Milky Way, neither the radio stars and pulsars, nor any of the myriads of visible stars play any significant part in forming this pattern, and we must look to the gas between the stars for the origin of the radio waves. Here we find electrons with cosmic-ray energies, and we meet a most important mechanism in the production of non-thermal radiation. This is the radiation from an electron with such a high energy that it is moving with nearly the speed of light. It can only radiate when its course is deviated from a straight line; in interstellar space the only agency for this is the magnetic field which permeates the whole of our galaxy.

Radiation from a high-energy electron in a magnetic field is usually known as synchrotron radiation, so called because it can be observed in the big particle accelerators known as synchrotrons. (It is also known as '*Magnetobremsstrahlung*', i.e. radiation from magnetic braking.) Galactic synchrotron radiation can tell us much about the number and energies of the cosmic-ray electrons, and about the strength and direction of the interstellar magnetic field.

Embedded in this bright band of non-thermal, synchrotron radiation are clouds of hot ionized gas surrounding the hottest and youngest stars. At short wavelengths these clouds, the H II regions, shine out over the non-thermal background like a string of pearls, while at long wavelengths where the synchrotron radiation is strongest they may by contrast actually absorb the non-thermal background radiation and appear as dark clouds against a bright background.

Although few of the visible stars can be seen with our radio eyes, while most of the observable radio emission is synchrotron radiation from interstellar space, these nebulous H II regions are closely linked in their radio and optical emissions. Other nebulae show more prominently in radio than in light. These are the supernova remnants, which shine with synchrotron radiation over most of the electromagnetic spectrum, but particularly brightly in the radio spectrum. The visible remnant of the supernova Cassiopeia A is so hard to see (Plate X) that the remnant was unknown until its discovery by radio astronomers, even though it is a young and vigorously expanding cloud, only just over 250 years old.

Spectral Lines

The radiation described so far is broad-band radiation, covering radio frequencies over the whole radio spectrum. At certain well-defined frequencies there is also the narrow-band characteristic radiation from some particular atoms and molecules. Neutral hydrogen radiates at 1 420 MHz; water, ammonia, methane and many other molecules all have one or more of these 'spectral lines' in the radio spectrum.

Visible spectral-line radiation is familiar in sodium and mercury street lights. The yellow sodium light is largely contained in a pair of spectral lines at 590 nanometres wavelength (1 nanometre = 10^{-9} metre), while mercury has dominant lines both in the blue and green part of the spectrum. A spectroscopic examination of these lines would show that they have a measurable width, which is determined by the temperature of the gas in the lamp.

In the same way a radio spectroscope can be used to analyse the spectral lines radiated by gas in the Milky Way. Obviously the radio frequency of the lines can be used as a sort of catalogue of the atomic and molecular species present, and the strength of the lines may be a measurement of the amount of each species. The story is, however, rather more complicated, since several spectral lines turn out to be much stronger and narrower than expected. Here is another of the surprises: the lines can be amplified in space in a celestial form of the amplifier known on earth as a 'maser'.

The hydrogen line (Chapter 8) has been extensively used in measuring the dynamics of the Milky Way. The precise frequency on which the line appears depends on the velocity of the radiating hydrogen. We shall see that the pattern of velocities can be used to draw out the spiral structure of the disc of our galaxy, which has previously been hidden to our optical telescopes by the absorbing dust in the Milky Way.

The hunt for new molecular species has concentrated on the dark nebulae in the Milky Way. These are dense clouds, heavy with dust, which appear to be breeding grounds for molecules some of which are sufficiently complex to be called the building bricks of life (Chapter 9).

GALAXIES

The distant nebulae were often described in astronomical works of the beginning of this century, but their actual distances were then dismissed merely as 'large'. Sir Robert Ball, for example, in *The Story of the Heavens* (1894) gives the spiral nebulae only a very brief mention in comparison with other topics, and says:

It is believed that some of these nebulae are sunk in space to such an appalling distance that the light takes centuries to reach the earth.

The true situation was grasped by Edwin Hubble, one of the greatest astronomers of recent years. His devotion to the problem of distant nebulae was combined with the most fruitful use first of the 100-in telescope on Mt Wilson, and later of the 200-in telescope on Mt Palomar. Hundreds of light-years may seem an appalling distance, but his new distances were many times beyond those of any galactic objects, and quite beyond any description by mere adjectives.

The 100-in telescope was completed soon after the end of the First World War, and during the 1920s Hubble revolutionized our concept of the universe by demonstrating the aggregation of stars into galaxies, and by showing that our galaxy was only one of innumerable 'island universes' which were so distant as to appear as small diffuse nebulae. In the constellation of Andromeda there is one such nebula which can be seen by the naked eye. Hubble was able to show that this nebula was a great star system like our own galaxy, at a distance from us of about two million light-years. The diameter of our galaxy is less than a tenth of this distance.

The 200-in telescope was first used in December 1947, although it was not until eighteen months later that all tests and adjustments were complete. It had taken twenty years to build; no fewer than eleven of these years were occupied in grinding the concave face of the mirror. The aspirations and dreams of one man, George Ellery Hale, who inspired the whole programme of construction of the 60-in and 100-in telescopes on Mount Wilson, have now produced a beautiful scientific instrument, satisfying to look at and deeply rewarding to use.

With this instrument, nebulae of the Andromeda type can be detected to a distance of about 1 000 million light-years, and in this distance there must be well over 100 million galaxies. Eddington, in his book *The Expanding Universe*, gave the following multiplication table, which is still the best guess we have at the statistics of the universe:

$$10^{11} \text{ stars} = \text{one galaxy}$$
$$10^{11} \text{ galaxies} = \text{one universe}$$
$$(10^{11} \text{ is one hundred thousand million})$$

It is not surprising to find a large proportion of astronomy concerned with studies of extragalactic nebulae or to find a growing number of radio astronomers with the same concern. The study of the distribution of the nebulae in space is the most fundamental of astronomical studies: it is the study of the structure of the universe.

Hubble photographed and classified a large number of extragalactic nebulae. They vary in shape from symmetrical, almost spherical assemblies of stars, through forms with a regular spiral structure, as in our own galaxy, to very open spirals and peculiar twisted and irregular shapes. The smooth nebulae are known as ellipticals, and are classified from E0 for a spherical shape, to E10 for highly flattened ellipses. The spirals are classified Sa, Sb, or Sc, in order of decreasing tightness of the spiral structure.

No nebula is completely described by this system alone, however, and the radio astronomer has often to deal with galaxies which do not fit into the sequence at all, and can only be described as 'peculiar'. There are, however, some new categories which have recently emerged, such as 'N' galaxies with condensed, brilliant nuclei, and 'dumbbells' with two nuclei in one envelope. These, along with the quasars, are often very powerful radio sources, and they deserve chapters on their own. Among the normal spiral and elliptical galaxies there are many which are normal in the radio sense, that is to say they behave rather as our own galaxy does. Some of these belong to a local group of galaxies, of which our own galaxy is a member, which seem to be associated locally in space, travelling together, and unaffected by the headlong recession of the more distant nebulae.

Several members of the local group will appear in our more detailed discussion of galaxies (Chapter 10). They have various descriptive names, such as the Magellanic Clouds, the Whirlpool, etc., but mostly they are referred to by the more prosaic catalogue numbers. The first catalogue of nebulae, compiled by Charles Messier (1730–1817), contains all varieties of nebulous objects, including supernova remains (M1, the Crab nebula), globular clusters of stars (e.g. M3, M5), and H II regions (e.g. M8), but mainly it is a list of extragalactic nebulae, even though Messier did not recognize them as such. For example the Andromeda nebula is M31, and M33 is another member of the local group. Other prominent spiral galaxies are M51, M81 and M101. We shall also refer to the elliptical galaxy M87.

Another catalogue, the New General Catalogue compiled by J. L. F. Dreyer in 1890, is often quoted. The Andromeda nebula (M31) is also NGC 224, and M81 is NGC 3031. But the NGC and its extensions (the Index Catalogue, IC, of 1895 and 1910) cover so many more nebulae that the fainter nebulae, which are often the most exciting, may only have an NGC number. The explosive galaxy NGC 1275, for example, has no M number, while the Cygnus A radio galaxy is so faint as to be unremarked and uncatalogued until its importance was revealed in 1952 by the accurate location of the radio source.

Radio Galaxies

The radio source in Cygnus was the first discrete radio source to be found. It appears on the map made by Reber, in 1944, of extraterrestrial radio waves at 500 MHz, but it was Hey who first recognized that it was a discrete source rather than a cloud of emission several degrees across. The accurate location of that source was a challenge taken up by radio astronomers in Sydney and in Cambridge. New aerials were built in both laboratories, and the position of the Cygnus 'radio star' was rapidly narrowed down to a small fraction of a degree. In that area, however, were many stars and extragalactic nebulae, and a casual glance at a photograph of the region revealed no object with any outstanding peculiarity to match a most remarkable radio emission. The only solution was to build a radio telescope which was specifically designed for measuring positions to the greatest possible accuracy, and this was undertaken by the Cambridge group. The experiment was rather like surveying, using a pair of exactly similar aerials at opposite ends of a 1 000-ft baseline. The position of its baseline, and its length, had to be obtained as a basis of the survey of radio source positions. Results, which included the position of the radio nebula in Cassiopeia (see Chapter 7), were sent in 1951 to W. Baade, who decided that the accuracy at last justified the use of the 200-in telescope. It is no use searching for a radio source with this instrument, whose plates cover a field of view only ten minutes of arc across, if it is necessary to make a dozen exposures each lasting many hours to cover the limits of error of the radio position. It would take up too much valuable telescope time for a quite uncertain result. However, the new position was believed to be accurate to within one minute of arc. On the first plate taken for this position, a nebula appeared at exactly the right place, and it was the most peculiar nebula ever observed.

Baade saw that this nebula, shown in Plate XI, had two nuclei, and he realized that they were so close together that if they were two galaxies, they must be undergoing a direct collision. As he had for several years been on the lookout for a collision between galaxies, he set out full of enthusiasm to demonstrate that this was indeed the first example of such a catastrophe. He discussed the possibility with his colleague R. Minkowski, who challenged Baade for proof from the spectrum of the nebula. A famous bet was made, with the stakes a bottle of whisky.

The spectrum turned out quite unlike any previously seen, with a series of emission lines which could only arise in gas in a very high state of excitation. Minkoswki accordingly handed over a bottle of whisky, and for several years the Cygnus A radio source was classified as 'colliding galaxies'. As we shall see, this interpretation was probably quite wrong, but the whisky had been drunk long before this was realized.

Cygnus A is now regarded as the prototype of some hundreds of radio galaxies, in which there may be many variations on the central theme which led to the discovery; they all are much stronger radio emitters than our own galaxy, by a factor typically of about one hundred thousand. There is also an important difference in the way in which the emission is distributed across the radio galaxies. As we shall see in Chapter 10, the radio emission no longer follows the same features as the stellar populations which determine the visual appearance of a galaxy; in Cygnus A, as in many others, the radio emission is from two locations right outside the visible galaxy. We can only speculate that these two emitters, one either side of the galaxy, have been shot out from the centre in some cataclysmic explosion.

Quasars

As the discrete radio sources were more accurately located, it became painfully obvious that radio galaxies alone could only account for about half of them. Photographs of the sky with large optical telescopes often show as many distinct galaxies as they do foreground stars in our own galaxies. The positions of some radio sources coincide with neither a star nor a galaxy. The catalogue entry 'blank field' covered a situation usually explained by an accidental visual faintness of a strong radio galaxy at a great distance. So when the only coincidence was with a star-like object, the same explanation was used on the assumption that the object was indeed an innocuous star in our own galaxy and not in any way connected with the radio source.

The first suspicion that not all radio sources were galaxies came from measurements of their angular diameters. This work, carried out by Henry Palmer at Jodrell Bank, involved the use of pairs of radio telescopes separated by distances of tens of kilometres, forming an interferometer (Chapter 16). The larger the spacing, the better the angular resolving power of the interferometer. It was natural to expect that the angular diameters would all be somewhat like those of the identified radio galaxies; depending of course on distance, these were expected to be of the order of 1 minute of arc. To everyone's surprise, a proportion of the radio sources had immeasurably small angular diameters, even when the interferometer baseline was extended to provide an angular resolving power well below 1 second of arc. Some of these radio sources were then found to be star-like objects which had previously been ignored. They were in fact not stars at all, but an entirely new type of extragalactic object, now known as 'quasars'. The story of these 'quasi-stellar' objects is told in Chapter 11.

The Depths of the Universe

The contrast between the radio sky and the visible sky is greatest and most significant in the realm of the galaxies. On the one hand the familiar nebulae, studied by Hubble and shown by him to be island universes, whole galaxies receding from us with enormous velocities, become for radio eyes so insignificant that only the brightest thousand or so can be seen. On the other hand there are at least a hundred thousand detectable radio sources, outside our galaxy, and reaching to distances beyond those available to Hubble. The key to this difference lies in an unknown process which turns a small proportion of galaxies into either quasars or radio galaxies – or possibly both, since there may be an evolution from quasar to radio galaxy. Whether this happens to most galaxies for a short portion of their total life span, or to only a few galaxies for a longer time, the effect is that radio eyes see a smaller population of galaxies, but they see them out to a greater distance. For cosmologists it is the distance that lends the enchantment to this view, since astronomical distances also mean astronomical ages.

The time of travel for light and radio waves from the most distant galaxies is measured in thousands of millions of years, a time scale which is comparable with the age of the universe. We therefore see, now, the distant parts of the universe in the state it was in during a much earlier phase of its existence. Cosmological history is spread out before us. The importance of the quasars and radio galaxies lies only partly in the astonishing physical processes occurring within them: it is their role as signposts to the structure of the universe that makes their discovery the most significant achievement of radio astronomy.

CHAPTER 4

The Sun and its Corona

Radio Waves from the Quiet Sun

ONE of the most startling and important features of the solar atmosphere is the extremely high temperature – about 1 million degrees absolute – which is maintained through the whole of the outer regions (the corona). This fact became an accepted one, on optical evidence, at about the same time as the first radio observations of the sun were being made.

The visible surface of the sun, called the photosphere, is at a temperature of 6 000 K, and is surrounded by the chromosphere, a region several thousands of miles thick, which is at about 20 000 K. No one would reasonably expect to find temperatures of a million degrees further away than these regions from the sun's centre. For the source of all solar energy lies in the very depths of the sun, where hydrogen is being converted into helium at the rate of 500 million tons per second, and heat flows out to the surface by radiation from atom to atom inside the sun. There is a steady gradient of temperature all the way from 10 million degrees at the centre to 6 000 K on the surface.

The temperature of the corona was originally measured by spectroscopy. Two methods were used, the first being a study of the Fraunhofer absorption lines which have already been mentioned in Chapter 1. The details of these lines show that the electrons in the corona, which scatter the photospheric light to make a visible corona, are in violent motion, giving the line a width which is increased by the Doppler effect. The velocities deduced from the width of the line correspond to the thermal motions expected from a corona at a temperature of 10^6 K.

In the second spectroscopic method some spectral lines were observed in light emitted by the corona, the wavelengths of these lines being characteristic, not only of the atoms, but of their state of ionization, which is in turn dependent upon the temperature.

In this dependence lies the solution to a long standing mystery of some of the so-called 'coronal lines', which do not correspond with any which could be produced in the laboratory. The mystery of these lines was at one time such that they were supposed to originate in atoms of a new unknown element, named 'coronium'. It has now been found that the coronal lines come from the highly ionized atoms of several quite ordinary heavy elements, such as iron in the form known as Fe XIV, which is an iron atom which has lost no fewer than thirteen electrons as a result of the high temperature of its surroundings. (Spectroscopists designate an un-ionized atom by the roman I, adding to this for each electron lost.) Accepting this evidence for the temperature, and other optical evidence for the density of the corona, we might fairly expect the sun to look like a large, blurred, million-degree radio emitter, at all radio wavelengths.

The first careful measurement of the temperature of the source of solar radio radiations gave an answer of 6 000 K, which is the photospheric temperature. This measurement was made by Southworth in 1942. Later the result was corrected to 18 000 K, but even this is much less than 10^6 K. The result occasioned no surprise at the time, for the high temperature of the corona was not yet widely accepted. Southworth's work was done on very short wavelengths, between 1 and 10 cm, and this proved later to contain the explanation of the low temperatures. Later work on longer wavelengths, particularly on wavelengths longer than about 50 cm (i.e. of radio frequency less than 600 MHz), gave results which, if expressed in the same terms, do indeed indicate temperatures in the region of a million degrees.

The graph in Figure 3 shows the variation in apparent temperature of the quiet sun over the radio spectrum. Broadly speaking, at the shorter wavelengths the temperature is near to that of the photosphere, and at the longest, to that of the corona. The difference shows a difference in origin, and the reason is not hard to find. The corona is completely transparent for short wavelengths, just as it is for light. Radio waves are generated in any hot body, and their intensities correspond to the temperature of the body. The photosphere, therefore, generates 6 000 K radio waves, and the chromosphere generates 20 000 K radio waves. If the chromo-

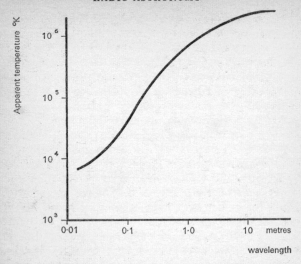

3. Graph showing the temperature of radio waves emitted by the sun. High-temperature radio waves are emitted by the corona at long wavelengths; low-temperature, shorter radio waves by the chromosphere and photosphere

sphere as well as the corona were transparent for radio waves, as they both are for light waves, the radio temperature would be 6 000 K. At the very shortest radio wavelengths, less than 1 cm, this might possibly be observed, but it is usual at short wavelengths to find that it is the chromosphere which is the effective radiator, with the photosphere hidden behind it. The radio temperature is then around 20 000 K. On long wavelengths, the corona not only plays the part of a screen which prevents radio waves from the photosphere and chromosphere from reaching us, but it also radiates its own million-degree radio waves.

If we are now to press home the analogy between seeing the sky with our new radio eyes, and seeing it in ordinary light, we must specify the wavelength of radio waves which our new eyes will admit, rather as we might specify the colour of the light admitted by a tinted pair of spectacles. Radio eyes working at a long wavelength would see the sun as a large flaming cloud completely obscuring the familiar sphere; for short-wavelength eyes the sphere is only slightly changed, being a little larger and marked with streaks and blobs corresponding to the well-known sunspots and other optical markings. At each wavelength the intensity of the radiation betrays the temperature of the region which is seen to be emitting.

The Brightness Distribution across the Sun

As well as the apparent variation of temperature of the sun with wavelength, we find a variation of the size of the source of solar radiations according to the wavelength at which measurements are made. Again using the idea of radio eyes, it is as though the source is large for red light and small for blue. The solar corona is several degrees across, and the photosphere only half a degree. Furthermore, the corona is not just one thin shell, but an electron cloud whose density falls steadily with height. As the density falls, the wavelength at which the corona can absorb radio waves increases. The absorbing sphere is not a single shell but contains layer upon layer of different densities and temperatures, rather like the skins of an onion. By observing on progressively shorter and shorter wavelengths, we may peel off the successive layers of the onion, examining each one before we go down deeper. At each stage the size of the sun will be different, depending on the wavelength. This possibility of exploring the sun's corona, by measuring the distribution of 'brightness' across the radio sun at a variety of wavelengths, presents a real challenge to the radio astronomer, offering him information that is practically denied to optical astronomers. In this way he may investigate the variation of both electron density and temperature with height above the photosphere. Let us see how the radio telescope is applied to this problem.

This measurement is not at all easy. Radio telescopes differ from optical telescopes in various rather obvious ways, such as the use of wire mesh instead of glass for reflecting radio waves instead of light. Two more fundamental differences are that the radio telescope cannot take photographs, and that the radio telescope has such poor angular resolving power that all but the more elaborate instruments cannot tell the difference between a point source of radio waves and a disc about one degree across.

The first techniques used for measuring the brightness distribution across the sun followed a scheme devised by Michelson for measuring the angular diameters of visible stars. This is the technique of interferometry, which now plays a more important role in the study of quasars (see Chapter 11). Extensions of this technique have been made, notably in Australia, so that a widely

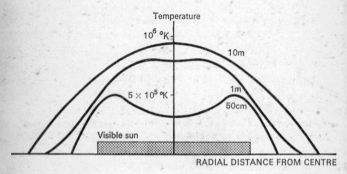

4. Radio-brightness distribution across the sun at long wavelengths. The longest wavelengths are generated far out in the corona, which is at 10^6 K; shorter wavelengths come from the cooler chromosphere just outside the visible surface of the sun

spaced array of small radio telescopes can be connected together to produce a single radio telescope with a high angular resolution. The best example of this is the Culgoora spectroheliograph (Chapter 17), which produced the picture of the active sun in Figure 14, using the long wavelength of 3·7 m. Typical cross-sections of the quiet radio sun obtained by interferometers at long wavelengths are shown in Figure 4.

At shorter wavelengths the bright corona is seen only near the edges of the solar disc, which is mainly seen as radiation from the cooler chromosphere and photosphere. This 'limb-bright' distribution is seen in Figure 5, which was obtained at a wavelength of 3 mm at an eclipse.

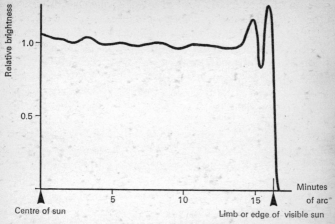

5. Limb-brightened distribution of solar radio emission at 3-mm wavelength. The limb-brightening in this eclipse observation shows an unusual doubling (after J. P. Hagen, Pennsylvania State University)

Solar Eclipses

Another method of measuring the radio-brightness distribution across the sun depends on the exciting but comparatively rare phenomenon of a solar eclipse. It is a remarkable coincidence that the moon and the sun subtend very nearly the same angle, $\frac{1}{2}°$, at the earth. Consequently when, as it occasionally happens, the moon comes between the earth and the sun, it just fits over the photosphere, affording a spectacular view of the light from the corona to those fortunate enough to be at the right place, at the right time, always provided that the weather is good enough (see Plate XIII). Expeditions to observe solar eclipses would not be nearly so important if the fit of the moon over the sun were less exact, which is the case for the radio sun. At long wavelengths the radio sun appears to be several degrees across, as it is the corona itself which generates the radio waves. It is only at shorter wavelengths, where the chromosphere or photosphere is concerned, that a significant eclipse can be observed. For wavelengths between 1 and 10 cm, the fit is not too bad, and the moon can cover up to about 90 per cent of the radio sun. The progressive drop in the radio waves reaching the earth as the moon covers up a larger and larger area delineates the details of the distribution of emission across the sun.

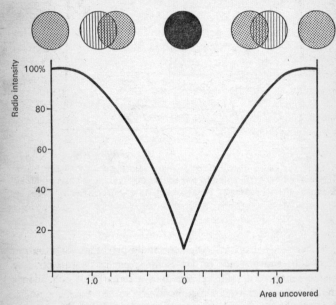

6. Radio eclipse of 30 June 1954. The eclipse was observed in Sweden by Mayer, Sloanaker and Hagen, using a wavelength of 9·4 cm. The curve shows the experimental result as the sun was progressively covered and uncovered by the moon

Some beautiful experimental results have been obtained in these radio eclipses, some when the sky was overcast, so that optical work was impossible. There are stories of astronomers who have attended as many as six solar eclipses, at all of which cloud spoilt their measurements. They must envy the radio astronomer his independence of weather. Figure 6 shows an eclipse curve made on the wavelength of 9·4 cm.

The first radio eclipse observations hold an interesting place in the history of radio astronomy. They were made by Russian scientists at the eclipse of 20 May 1947. Two main problems of eclipse observations, the transport of rather large aerials to the observing site, and the directing of this aerial towards the sun, were solved by the masterly stroke of installing the whole apparatus on shipboard. The aerial consisted of ninety-six dipoles over a reflecting sheet, and as it was fixed to the ship, it was directed by hauling on the mooring lines of the ship. A wavelength of 1·5 m was used, at which the radio sun is not covered very efficiently by the moon. The results were not, therefore, of great precision, but the experimenters were able to suggest that irregularities in emission over the sun's surface were associated with some specific optical features.

The Quiet Sun and the Sunspot Cycle

Photographs of the solar corona taken at times of different eclipses show a large variation in the shape of the corona. At times when the sunspot activity is least, such as the eclipse of June 1954, the corona extends out in long streamers in the equatorial plane, being visible to distances of about five times the solar radius. At the poles, a different type of streamer can be seen: distinct lines or rays which seem to follow lines of magnetic force, as do iron filings around the pole of a magnet. At times of sunspot maximum, however, the corona is more nearly the same around the whole disc, although it is often distorted with streamers going off at odd angles. This difference in behaviour follows the well-known 'sunspot cycle', in which the numbers of sunspots to be seen on the sun, and their size and position, vary in a fairly regular pattern through a cycle of eleven years' duration. The differences in the corona are so striking that we would expect to see considerable differences in the quiet-sun radiation through the cycle.

The first regular recordings of the solar radio waves were started in 1946, so that they now cover two complete sunspot cycles of eleven years. Variations are certainly found, but it is hard to say whether the whole basic level of radiation changes, as the extra radiation coming from obvious disturbances on the sun masks any variations of this basic level; in other words, the sun is never really quiet during 'sunspot maximum' years. This extra radiation comes not only from the spectacular disturbances on the sun, noise storms, and flares (discussed in the next chapter), but seems to be associated with the regions rather curiously known as 'calcium plages'. These are areas of optical disturbance marked by the unusual emission of spectral lines from calcium, and are a common and long-lived feature of the sun. The long series of recordings gives us good ground for associating them with radio emission, particularly on wavelengths of about 50 cm. There are large cyclic variations of both the radio emission and the exposed area of calcium plage as the sun rotates, and the variations match one another very closely.

The Solar Corona

The difficulty of explaining the steady maintenance of the solar corona in its present state is considerable. Firstly it has to be heated to a million degrees, from a source of energy whose surface is only at 6 000 K. Secondly it must not 'boil off' faster than it can be replenished. 'Boiling off' occurs when the most energetic of the particles in the corona have sufficient velocity to escape from the sun's gravitational field. Our own atmosphere is boiling off slowly, although the common gases are not much affected; the lighter gases tend to be lost faster, otherwise helium would certainly be a common gas. The moon, being lighter and hence exerting less gravitational pull, can retain no atmosphere at all. The sun has a gravitational force at its surface twenty-seven times that of terrestrial gravity, but the corona is three thousand times as hot as our atmosphere and extends far out into space, so that boiling off is nevertheless quite possible. Energy for heating the corona could have come from this strong gravitational field, as matter could be attracted to the sun and fall into it with high velocities. This is called the 'accretion' theory of heating. A more generally accepted theory at present is that the energy comes from the high-temperature regions inside the sun, and is transmitted not directly as heat, but as mechanical oscillations. These waves are rather like sound waves, but involve also the magnetic field of the sun. The waves are called magneto-hydro-dynamical waves, in which the magnetic field acts on the ionized gas rather as though the gas were held together with elastic cords. In this theory, the waves are generated below the photosphere and pass out into the corona, where their energy is dissipated when the density becomes too low to support them. Qualitatively, this argument is plausible; quantitative answers are rather elusive, and the reason for a coronal temperature of a million degrees, rather than say 100 000 K, is unexplained. One interesting facet of the problem is that the corona finds it rather difficult to get rid of its energy, since it radiates very little, and it cannot easily conduct heat down into the cooler photosphere. Outside the sun there is nowhere for it to conduct heat to at all. There must, however, be a balance sheet of energy gain and loss, and it appears likely that one of the most important losses is by the same radio radiation which tells us of the existence of this thin hot mantle of gas round the sun.

From the radio emission of the solar corona, a useful picture of the electron density and temperature can be made out to distances of about three or four times the radius of the sun. Beyond this distance the corona must be studied in different ways: indeed it assumes a rather different character, that of an inter-planetary medium, extending out to such distances that we must think of the planets ploughing through a hot ionized sea of gas as they follow their orbits round the sun.

The Solar Wind

As far as ordinary appearances go, the earth is separated from the solar corona by 140 million km of empty space. Early work in geomagnetism first showed that there might be a stream of particles crossing this space, but only on the occasions when solar flares launched the energetic particles which were supposed to account for magnetic storms on earth. The existence of the hot corona implies that hot ionized gas is leaving the sun all the time, and shows that these 'storm' streams are only a perturbation in a steady outward flow, known as the solar wind.

The corona at its faintest appearance on eclipse photographs has a density of about 10^5 protons and electrons per cubic centimetre, i.e. 10^5 cm^{-3}. By terrestrial standards this is a very good vacuum; compare the density of 10^{13} atoms cm^{-3} in a radio valve, or 10^6 electrons cm^{-3} in the upper part of the ionosphere. There is practically no hindrance to the outward flow of the hottest part of the coronal gas beyond the visible limit of the corona, so that the density falls off smoothly and comparatively slowly, and the speed of the solar wind remains high throughout interplanetary space.

The solar wind has been observable, although not recognized, from time immemorial in its effects on the tails of comets. These tails are blown by the wind to point radially out from the sun, acting as weather-vanes. Direct observation of the solar wind was first achieved by a Soviet spacecraft in 1959, and subsequent space probes from USA and USSR have provided considerable details of its characteristics. For example, the density at the distance of the earth is between 5 and 50 electrons and protons cm^{-3}, and their velocity is between 400 and 800 km sec^{-1}. The wind carries with it a magnetic field, drawn out from the general magnetic field of the sun like strings of spaghetti; the field amounts to between 5×10^{-5} and 20×10^{-5} gauss, which is of course a very small field compared with the terrestrial magnetic field. The solar wind, and its magnetic field, are both variable in speed and direction; they are gusty, and the field lines lash about like the stream lines in a river.

Direct observations of the corona by spacecraft offer only a very small sample of conditions. It is, for example, very difficult to place a spacecraft orbit at an appreciable angle to the plane of the ecliptic, in which the planetary orbits all lie. It is also very difficult to arrange to fly a spacecraft close to the sun, since the spacecraft might overheat and share the troubles encountered by that early space explorer, Icarus. In both these regions radio astronomy can provide some useful information, though with rather different techniques than the observation of direct radiation which has been described so far.

Radio Occultations

Radio emission from the corona can give a good picture of the density and temperature out to a distance of about three solar radii. (The sun's radius is 700 000 km, or 450 000 miles.) Beyond that, the density is so low that any radio emissions would not be detected on earth. They would be of a very long wavelength, and would not reach us because of the blanketing effect of the iono-sphere.

Radio emission is, however, not the only way in which the corona shows itself to radio astronomers. It can refract and scatter radio waves, just as a turbulent cloud of hot air can disturb the view of objects beyond it. Hot air does not emit light, but it can refract it, and it can be seen by its effect on the visibility of distant objects, which become blurred and distorted as eddies of hot air cross the field of view. Layers of warm air over sunlit sands, road surfaces, or over the sea in calm conditions can so refract light that things appear to be in quite the wrong places, ships appearing to be in the sky, and the sky appearing on the road or desert surface as if reflected in pools of water. The corona is a region where the refractive index for radio waves can be significantly less than unity, which is its value in free space. Where it is zero, radio waves can be generated; where it is not zero, but is varying and slowly approaching unity as the distance from the sun increases, a bending of radio waves should take place.

Figure 7 shows the effect of this refraction on a radio wave of about 4-m wavelength. The sun may be seen to present a disc of diameter about three times the visible disc as an impassable barrier for these long radio waves. At lower frequencies this occulting disc is even larger. A radio source which happened to be behind this disc would be completely cut off, and its radiation (as seen from the earth) would disappear. It would not be detectable again until the motion of the sun carried the corona several degrees across the sky, so that the radio source might disappear for as long as three days if it happened to be exactly on the track of the centre of the sun, in its path round the plane of the ecliptic.

It so happens that one of the brightest of radio sources, the Crab nebula in Taurus, lies very close to the plane of the ecliptic, so that in June of each year the sun passes within about $4\frac{1}{2}$ solar radii from the path of the radio waves travelling from the Crab nebula to our radio telescopes. In Cambridge, in June 1950 and again in 1951, several instruments were set to watch this radio source day by day, with the hope that its radio waves would disappear behind the furthest parts of the solar corona, on or around

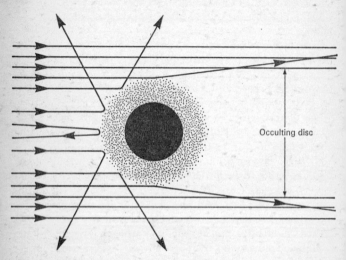

Occulting disc

7. Bending of 80-MHz radio waves by the solar corona. The occulting disc, through which radio waves cannot pass directly, is much larger than the visible sun

14 June. Nothing but frustration came from the experiments of these two summers, as the sun was so active that on every day in the middle of June intense sunspot radiation swamped the receiver, completely masking any effects on the Crab nebula. This was also the experience of Russian radio astronomers, who were quite independently trying out the same idea.

In June 1952 the experiment was repeated successfully – with greater success, indeed, than could have been expected. Instead of losing the Crab nebula for one or two days as expected, and that on the lowest frequencies only, it was found that the Crab was being affected by the corona for over a week, on all three frequencies in use. This meant that the corona was observable out to 10 solar radii. The observations of the occultation were all made with the use of interferometer aerial systems, which are sensitive to the angular size of the radio source (see Chapter 16), and it turned out that what had been seen was not a regular refraction, but an irregular diffraction – in other words, a heat haze, and not a mirage (see Figure 8). The corona must contain irregular clouds

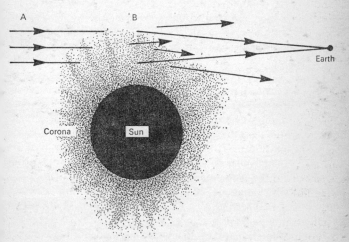

8. Diffraction in the solar corona. Parallel rays from a distant radio source (A) are scattered in the irregular corona (B) and are received over a considerable angle at the earth

of electrons, through which the small bright radio sources become diffuse sources of radio waves, not losing energy by absorption in the corona, but nevertheless recording lower values on the interferometers, because of their distorted shape.

Scintillations

The blurring effect of the solar corona on a small radio source is so like the effect of a heat haze that we might expect to see another familiar effect, scintillation, or 'twinkling'. Stars twinkle, but planets generally twinkle rather less, on account of their greater angular diameter. The atmospheric irregularities which are responsible for scintillations are generally a few inches across, and as they blow across the sky a star appears to move about irregularly by a second of arc or more, rather too fast for the detailed motion to be followed by the eye. At the focus of a telescope a fixed photo-electric tube with a small aperture would give a fluctuating output as the star image danced by: this is 'amplitude scintillation'. A photographic plate exposed for several seconds would only show the blurred image covering the whole area of the dance.

Amplitude scintillation can be observed in the radio signal from a distant source seen through the irregularities of the solar corona provided that the radio receiver can respond fast enough, and also provided that the radio source has a small enough angular diameter. At about 300 MHz the scale of the irregular pattern of radio waves on the ground, which is like the irregular pattern of light on the bottom of a swimming pool due to the waves in the surface, is about 200 km across. It sweeps past at the speed of the solar wind, which is the speed with which the outer part of the corona is flowing out from the sun. The solar wind flows past at about 300 km sec^{-1}; the fluctuations therefore last about half a second.

Observations of scintillations evidently refer to the irregular part of the far corona; they do not give any value for the average density, which is only measured by spacecraft. The present state of knowledge on these irregularities may be summarized as follows:

(i) They are present through the whole of the eleven-year solar cycle, but they are more noticeable at sunspot maximum.

(ii) They extend through the whole of the solar corona, including the polar and equatorial regions, and out to distances beyond the earth's orbit.

(iii) They are the smallest known feature of the corona, with a size everywhere less than 1 000 km.

(iv) They represent about 10 per cent of the total electron content of the solar corona.

It is remarkable that such a feature of the solar corona should not be observable in any other way than through radio scintillation. Nevertheless the effect at long radio wavelengths is very obvious; for example a quasar observed close to the sun at a wavelength of 10 m would be so blurred by scintillation that it would appear to be a degree or more in diameter rather than its true angular diameter of less than 0·01 arc seconds.

CHAPTER 5

Sunspots, Flares and the Earth

SIR EDWARD APPLETON occupies a very special place in the history of scientific radio. He was the first man to prove the existence of the ionosphere by demonstrating the reflection of radio waves, and the E-layer of the ionosphere, which he discovered, is often named after him. He continued with ionosphere research for over forty years since that famous discovery, and with his energetic exploration of many new lines of research it is not surprising to find him associated with the earliest observations in radio astronomy. Characteristically, his part was to realize the importance of a little publicized observation, and to demonstrate its physical significance.

The observations were of a strange sort of interference, experienced by radio amateurs experimenting with short-wavelength communications. The interference was of a subtle kind, since it gave very much the same effect as a deterioration in the performance of their receivers. Any radio receiver has a level of sensitivity below which it cannot receive signals, and this level is determined in good receivers by the electrical signals generated in the first valve amplifying stage. These signals sound like a hissing noise, and are called by radio engineers quite simply 'electrical noise'. No receiver is free from it; a deterioration in the performance of a radio communications link may well mean that the receiver has become unduly noisy. Jansky was very much concerned with just this noise in his original work in 1930–32, although he was in fact recording the *extra* noise, of the very same character, that is received as radio signals from our galaxy.

The amateur work was done in the middle of the 1930s, several years after Jansky. The sky noise was found on occasion to increase to such a degree that the galactic signal itself was completely swamped, and even strong communication radio signals were lost in the extra noise. One can imagine the 'hams' of those days, checking every single stage of their equipment to track down this extra noise and finally concluding that it must be just one of those awkward things that are sent to try us. But radio amateurs have always had a reputation for serious and painstaking research; they compared notes, collected times and intensities of signals, and looked for an explanation. Finally D. W. Heightman concluded that the extra noise was associated with solar activity. This conclusion explained the mysterious fact that Jansky had seen nothing of this trouble. For the sun was known to have a periodic change in activity through a cycle lasting about eleven years, and Jansky's experiment in 1931 had been at a time of minimum solar activity.

Appleton's attention was drawn to this phenomenon, but he was already being swept, as were many other radio scientists, into the feverish activity of developing Britain's new radar defence system. No further results were reported, the solar cycle waned, and the subject was dormant, and almost forgotten until it was dramatically awakened in 1942.

When the war started in 1939, a radar screen had already been constructed around our coasts, a screen which made possible the victory of the Battle of Britain. Many other radar devices were being developed which were later to play important parts in the war. Amongst these were the radar sets used to guide and control anti-aircraft guns, the gun-laying, or G.L. radars. These sets were to be found all over England, sending radio pulses into the sky, with sensitive receivers waiting for the message that a returning echo would give. The direction and range of an enemy aircraft would then be obtained immediately, and the guns could go into action. Of course, the enemy knew about this, as we knew of his radar systems, and a grim game was played throughout the war of jamming or confusing one another's radar systems by various types of radio counter-measures.

To jam a radar system is easy. The enemy aircraft, wishing to avoid the accurate plotting possible on a radar system, transmits a powerful signal on the radar frequency, which swamps the receiver and makes the normal tracking system useless. But of course the receiver can still determine the direction of the aircraft from the direction of the jamming signal. A far more effective jammer would be one which could put a radar set out of action without the operator realizing that anything was happening. It is possible to do this by transmitting electrical noise, which will deceive the radar operator into thinking that his set performance is rather poor at that time.

Not all operators accepted this easy explanation, when on 26 February 1942 they found the noise level of their sets had practically hit the roof. They soon found the noise to be coming in through the aerials, not generated in the receiver itself, and many were able to give accurate 'fixes' on the source of the noise signals. At the Army Operational Research Group, Dr J. S. Hey collected all the reports together. His aim was to find out the location of the noise transmitter, and to find some way of preventing the G.L. radars from being put out of action every time the enemy turned it on. He very soon found that the 'enemy' in this case was common to both sides; it was in fact the sun. The German radar systems must have been subject to just the same mysterious type of jamming, but both countries would naturally keep their reports secret. Hey's findings were at first known only to the radar research establishments and to the services, but they were properly published when the war was over.

Appleton, who was now able to show that the noise which the amateurs had received on their sets came from the sun itself, which was radiating with fantastic strength at these times, made the first announcement. Later a joint paper, by Appleton and Hey, associated the strongest bursts of radiation with solar flares.

The quiet-sun radiation at metre wavelengths is over a hundred times greater than that expected from a body like the visible sun at 6 000 K. This, as we have seen, arises because the sun is surrounded by an invisible corona at a temperature of about one million degrees. No such explanation could be given for this new enhanced radiation, as the intensity was of the order of a million times greater even than the quiet sun's signals. Later it was shown that only a small part of the sun's surface was now acting as a transmitter, and this area must have been so active that the idea of a temperature became almost meaningless.

The Origin of Sunspot Radiation

When the war ended, it again became possible for scientists to choose their own field of research, and several chose immediately to follow up the problem of this strange radiation from the sun. In Australia, a team under the head of Dr J. L. Pawsey, and in Cambridge a new section of the Cavendish Laboratory, under M. Ryle, both started to record solar radio emission on metre wavelengths, with apparatus of great sensitivity. They found that they were able to detect emission at all times, and started the earliest series of recordings of the daily levels of radiated power. They soon found that the power was very variable, with emission occurring sometimes for several days at a level some hundreds of times greater than that of the quiet sun. An outburst of this kind came to be known as a 'noise storm', and was found to be associated with visible sunspots. The largest increases were caused by solar flares, in which visible prominences are thrown out from the sun with incredible explosive violence. These outbursts were followed by the largest noise storms, and must have caused the noise interference to the earlier radio and radar observers. The key to the phenomenon lay in the sunspots.

A visible sunspot (Plate XIV) is a dark region of the sun's surface, covering an area which may easily equal that of Europe. Sunspots are dark because they are cooler than the rest of the surface; in large spots the temperature falls below 4 000 K, in contrast to the normal 6 000 K photospheric temperature. The root cause of the spot is the appearance at the surface of a very strong magnetic field. This field is in the form of a loop, which pushes up from a continuous belt of field below the surface. Sunspots are often bipolar, with separate regions where the loop leaves and re-enters the surface. At the surface the field may exceed 1 000 gauss, in contrast to the ordinary quiet-sun magnetic field which is about 10 gauss, about five times the earth's magnetic field.

The magnetic field of a sunspot extends high into the solar corona, and profoundly affects the emission of radio waves. New mechanisms of emission come into play, which produce strong non-thermal radiation. The importance of the magnetic field is

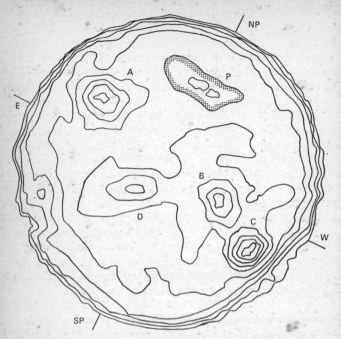

9. Map of solar radiation at 1-cm wavelength. The sun is seen as a disc of thermal radiation at a temperature of 7 000 K, with active regions A, B, C, D and a cooler region P (after Kundu and McCullough, Naval Research Laboratory)

demonstrated by the predominance of circular polarization (see page 27), which is only expected when a strong field forces the radiating electrons to follow a circular path. Figure 9 shows sunspot regions on a map of solar radio emission at 1-cm wavelength.

Outbursts and Flares

If we study the photosphere close to a sunspot with a spectroscope, which can isolate the light emitted by different elements, we find small bright regions emitting a hydrogen line (known as Hα) and a calcium line (Ca–K). These regions move about, and their brightness fluctuates as the spot grows and decays. Suddenly, within a few seconds, many of these bright spots coalesce and the whole area around the sunspot shines out brilliantly in Hα and Ca–K light. This is a solar flare. Sometimes the flare expands out into an enormous glowing arch prominence, as in Plate XV.

The flash of light from a solar flare represents an enormous

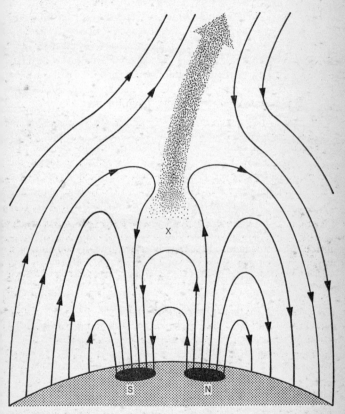

10. The magnetic field over a pair of sunspots. The two spots have opposite polarities. At the point x the magnetic-field lines can collapse, generating strong electric fields. Solar flares develop at this point, and electron streams propagate out along a magnetic channel, generating radio outbursts

release of energy, derived from the collapse of part of the strong magnetic field of the sunspot. In Figure 10 a line of magnetic field is shown which can switch to a shorter path, so that within the centre of the loop there can be a very sudden change of magnetic field. This has the immediate effect of accelerating all charged particles within the volume, in a very similar way to the operation of some high-energy particle accelerators in nuclear-physics laboratories. These particles, mostly electrons and protons, are then extremely energetic, with well-organized motions. The consequences of this explosive acceleration are spectacular and complex, affecting not only the immediate region of the flare but the whole extent of the corona above the sunspot, right out into the interplanetary gas. Furthermore, radiation of ultra-violet light, X-rays, and charged particles from the flare can reach the earth and profoundly affect the ionosphere. We will follow these events in time sequence.

The first and simplest result of the flare is an intense local heating of the lower part of the solar atmosphere, bringing the temperature up to over 10 million degrees, which is well above the coronal temperature. This hot region then radiates thermal radiation in the short-wavelength end of the spectrum, i.e. X-rays and ultra-violet light.

More complicated effects come from the electrons accelerated in the collapsing field. These have energies of up to 100 kilovolts, which may enable them to burst out of the flare region and travel at high speed into the corona. Others, however, are trapped in the remaining magnetic field, where they may radiate radio waves as they gyrate round the field lines. Only the shortest radio wavelengths are detected on earth at this time, since the corona continues to act as an impenetrable shield for longer wavelengths. The progress of the flare so far is illustrated in Figure 11 which shows the sudden growth and decay of ultra-violet and X-ray radiation as observed from a satellite above the earth's atmosphere, with the growth and decay of radio waves at 2-cm and 21-cm wavelength.

The longer radio wavelengths are emitted in an entirely different way. As the burst of electrons penetrates through the corona it passes through regions of successively lower density, each of

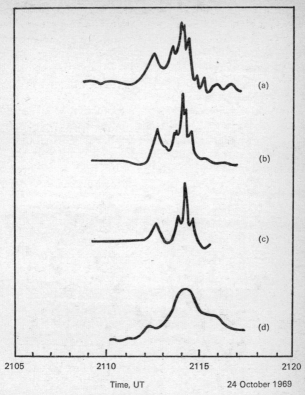

2105   2110   2115   2120

Time, UT   24 October 1969

11. A solar flare observed by X-rays, 3-cm radio, ultra-violet light, and as a disturbance of the terrestrial ionosphere: (a) X-ray flux recorded by the OGO 5 satellite; (b) 3-cm radio flux recorded in Cambridge, Mass.; (c) Ionospheric movement, from the reflection of a 10-MHz radio wave, recorded in Hawaii; (d) Ultra-violet light outburst, recorded by the OSO 6 satellite (from Harvard Observatory)

which can radiate a particular radio wavelength, that increases as the density decreases. These are the wavelengths of the quiet-sun radio waves described in Chapter 4. Each region is excited in turn by the burst of electrons; furthermore the organized electron stream forces it into very strong non-thermal radiation, so that a powerful burst of radio waves is observed with a steadily increasing wavelength, marking the passage of the electron cloud. The observed wavelength reaches the limit of earth-bound radio telescopes, at about 10-m wavelength after only a few minutes, when the electron cloud has reached a distance of several solar radii above the surface. The radio burst has, however, been observed at wavelengths up to 500 m by using radio receivers on satellites well above the terrestrial ionosphere; the low densities corresponding to such long wavelengths are so far from the sun that it is known that some electron clouds continue out through the corona at least as far as the earth's orbit.

The rate at which the wavelength of this radio burst increases can be used directly to measure the speed of the cloud; this turns

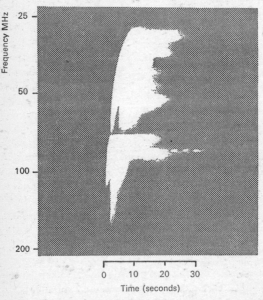

12. The dynamic spectrum of a solar outburst. The peak radio emission starts at about 150 MHz, and falls rapidly to 25 MHz during 10 seconds. This represents a movement of almost 100 000 km at one third of the speed of light, typical of a Type III burst. The emission at each frequency lasts for about 15 seconds (after J. P. Wild)

out to be typically one third of the velocity of light. These travelling clouds were discovered by the radio astronomer J. P. Wild, who set up in Sydney a receiver which swept continuously in frequency, recording the frequency at which the signals were strongest. Figure 12 shows a typical recording of one of these radio bursts, known as the Type III bursts.

Following the rapidly travelling burst, a slower disturbance, known as a Type II solar radio burst, travels out, again triggering off a radio burst at successive layers of the corona; this time the burst is stronger and longer-lived, and it also includes a component at double the main frequency. This can be seen in Figure 13, which is derived from one of Wild's observations. The location of the source of the radio waves far out in the corona is shown in the maps of Figure 14, which show the radio emission at 80 MHz.

Again following the passage of this second type of burst, there remains radiation from the unusually hot corona above the spot, persisting for some hours. This can hardly be said to complete the story, however, as there is a complex collection of smaller radio bursts described in the literature, some with frequency falling with time, some rising with time. As might be expected, these are quite different for the short wavelengths, which refer to conditions close to the visible spot, and for long wavelengths, which are concerned with the corona. Altogether, the radio emission from a solar flare represents a first-class solar firework display.

Effect on the Terrestrial Ionosphere

The effect on earth of a solar flare is far greater than the arrival of these unusual radio waves. As far back as 1740 it was known that compass directions occasionally suffered sudden deviations of many minutes of arc, which are now known to be associated with solar flares. There is a more regular compass deviation in the form of a daily cycle, swinging about 7 minutes of arc (in England) at times of sunspot minimum and about 11 minutes of arc at times of sunspot maximum. The connection between both of these compass deviations and the sun is through the ionosphere, which shows the most profound effects of a solar flare.

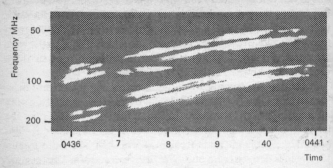

13. The dynamic radio spectrum of a Type II solar outburst. The two bands represent radio emission starting near 100 MHz and falling to near 50 MHz, together with harmonic radiation falling from 200 MHz to 100 MHz. This is a Type II solar radio burst, recorded at Culgoora Observatory on 9 October 1969 (after J. P. Wild)

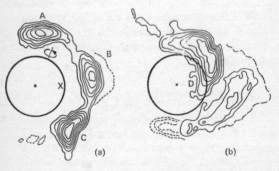

14. Solar radio outburst of 30 March 1969, recorded at 80 MHz by Culgoora Observatory. (a) The radio emission from the three regions A, B, C, high up in the solar corona was excited by a solar flare on the limb of the sun at X. (b) 15 minutes later the pattern has changed, with new flare emission from near the limb at D, and part of the older regions still emitting further out from the sun (after J. P. Wild)

The terrestrial ionosphere is an ionized region of the atmosphere, generally extending between heights of 100 km and 500 km. It can reflect radio waves of medium and long wavelengths, and so it is important in communication over long distances where a direct line of sight is prevented by the curvature of the earth. Without it, Marconi would not have received his radio signals across the Atlantic; his experiment was in fact an act of faith, since the existence of the ionosphere was almost unsuspected at the time.

The ionosphere is formed mainly by the action of ultra-violet light from the sun, ionizing the oxygen and nitrogen of the atmosphere. The free electrons resulting from the ionization can reflect radio waves with wavelength greater than a limit which depends on the number of electrons in unit volume, i.e. the number density. Thus, the lowest parts of the ionosphere, at 100 km and below, reflect the longest wavelengths, while shorter wavelengths penetrate higher in the ionosphere where the number density is greater. The ionosphere does not extend far below 100 km because the ultra-violet light is absorbed by the atmosphere at greater heights; furthermore ionospheric electrons do not remain free for long but are recombined to form neutral atoms, a process which occurs with increasing rapidity below 100 km.

This natural mirror in the sky, which has been an essential element in long-distance communication, does have its defects. One of these showed up during the sunspot maximum of 1937, at the same time that Heightman was receiving radio signals direct from sunspots. Long-distance radio reception, using ionospheric reflection, was suddenly cut, and remained a blank for several hours. The general public was made aware of this by a catastrophic failure of the Australian Test Match commentaries. Had the ionosphere disappeared?

The 'sudden ionospheric disturbance', or SID, as this phenomenon was named, is now known to be a dramatic increase in ionization of the lowest parts of the ionosphere. Normally the radio waves most used for long-distance communication penetrate these layers and are reflected higher up. But if the lower layers become very dense they have an extra effect of absorbing the radio waves as they pass through; the mirror becomes tarnished. This is the result of a solar flare. The extra ionization comes from the hot chromosphere, which is so heated by the collapse of the magnetic field above the spot that it can emit X-rays, which are even more potent than ultra-violet light in ionizing the atmosphere.

The magnetic deviations can now be understood in the same way. The regular diurnal swing of the compass needle is due to electric currents in the ionosphere, which ebb and flow as the ionosphere grows and decays through the day and night. The sudden deviations are the result of electric currents which are set up in the new ionization due to the pulse of X-rays at the time of the solar flare.

With the advent of submarine cables and satellite radio links, world-wide communications are no longer completely dependent on high-frequency radio, and at the mercy of solar flares. But high-frequency radio is still very much used, and there is considerable commercial value in any prediction which might be made of the incidence of solar flares and their accompanying communications blackouts. Another important requirement concerns high-flying aircraft, in which the passengers may be exposed to the particle radiation from a solar flare unless the aircraft descends to a safe height where it is shielded by the atmosphere.

There does seem to be some hope that solar flares can be predicted from the solar radio emission at short wavelengths, near 1 cm. This radiation originates near the surface, and is very sensitive to small changes in the magnetic field. There are often small radio bursts preceding the main solar flare, and if a clear enough pattern of behaviour can be established it should be possible to give a warning of the start of a solar flare, possibly about an hour in advance of the most serious effects.

CHAPTER 6

Supernovae and Radio Stars

The Crab Nebula

On 11 July 1954, a gathering of radio astronomers met in Cambridge to celebrate a birthday. This was no ordinary birthday party: it was the 900th anniversary of the birth of a new nebula.

In the ancient records of the Chinese astronomers appear several accounts of the appearance of bright new stars in the sky. For example:

2nd cyclical day, 5th month, 1st year of Chih-ho of Sung, guest star appearing at South East of T'ien-kuan several ts'un long, lasting more than one year . . .

and again:

60th cyclical day, 10th month, 2nd year of Chung-p'ing of later Han, guest star appearing at Nan-mên, big as half a mat, five colours and different tempers, later a little diminishing, disappearing in the 6th month, next year.

The first of these refers to the year A.D. 1054, a classic date for astronomy. It is possible, from this quaint account, to find quite accurately the time and place of one of the most dramatic events that man can hope to see – the supernova explosion of a star. No star was actually born at this time, but a star which was already old was suddenly transformed into one of the brightest objects in the sky. The star had been shining steadily, obtaining its energy from the conversion of hydrogen into helium, a controlled and orderly version of the hydrogen bomb. In the process it had been shrinking and becoming hotter inside. A core of helium was collecting, an apparently inert ash from the nuclear burning of hydrogen. But the temperature continued to rise until a critical level was reached, when helium itself began to release energy by combining to form nuclei of larger atomic weight. The succession of nuclear fusion continued through carbon right up to iron, until energy was released so fast that the whole massive core of the star exploded with the violence of a colossal thermonuclear bomb. The star was blown apart, as a glowing sphere of gas, leaving behind only an insignificant remnant. This sphere of gas is still to be seen with a telescope of moderate power, still expanding and glowing with light nine hundred years after the explosion; it is known as the Crab nebula in the constellation of Taurus (Plate VIII). It is the first in the list of nebulae compiled by Messier in 1784, and it is often referred to as the nebula M1. It is the only clearly recognizable remains of a supernova explosion, and as such it has been closely studied. It is fortunate that so much light is still emitted from the nebula so that we are able to observe it in great detail.

A radio astronomer who particularly should have been at the party is John Bolton, as it was he who made the discovery from the Sydney Radiophysics Laboratory in 1947 that the Crab nebula is a very powerful radio transmitter, nearly the strongest in the galaxy. At that time radio astronomy consisted of a rapidly growing study of solar radio waves, together with the old observations by Jansky and Reber of the background of cosmic radio noise. Bolton tried a new idea, and began to look for individual discrete sources in this background. The results of this search are another story, except that one of the first 'radio stars' to appear in his records turned out to be in the constellation of Taurus, and near enough to the Crab nebula for a suspicion, almost a certainty, that the Crab was the source of the radiation he had detected.

There have been other supernova explosions. The Danish astronomer Tycho Brahe, known for his accurate observations of the positions and movements of planets, and Kepler, the interpreter of these observations in terms of elliptical orbits, recorded the positions of bright new stars which for a time completely dominated the sky. Both of these men worked before the invention of the telescope, but nevertheless their positions are reliable to about 1 minute of arc. Very little is to be seen in these positions now, except for traces of wisps of gas, still blowing away. And even these were found only recently, with the help of the 200-in Palomar telescope. We cannot find out very much about these supernovae, as compared with the Crab nebula. Even their distance is uncertain, whereas the distance of the Crab nebula is known with greater certainty than that of most astronomical objects. In terms of galactic distances they are probably fairly local, and are to be found in, or near, the same spiral arm of our galaxy as our own solar system, but partly hidden from us by the obscuration of dust clouds.

The distance of the Crab is found by observing its actual rate of expansion in two separate ways. Firstly, the angular rate of expansion, as seen from the solar system, is measured from the changing appearance of the nebula in photographs taken several years apart. Secondly, the light from the centre of the nebula is analysed spectroscopically; it originates both in gas that is moving towards us at the front of the shell and in gas receding from us at the rear, and this range of velocities reveals itself in a Doppler spread of spectroscopic lines. The distance is found by dividing the expansion velocity by the angular rate of expansion, and is found to be 5 000 light-years. Light and radio waves therefore take longer to reach us than the actual life span of the nebula since the explosion was observed.

A Hidden Powerhouse

The Crab nebula has proved to be so rich, both for observers and for astrophysicists, that a distinguished theorist, Geoffrey Burbidge, has divided astronomy into two categories – the Crab nebula and the rest of the universe. Radiation from the nebula has been detected over the whole available electromagnetic spectrum, from radio through infra-red, optical, ultra-violet, X-rays to gamma-rays. This unparalleled performance immediately suggests a powerful, non-thermal source for the radiation; there is ample proof for this in another characteristic of all radiation from the Crab, which is the high degree of linear polarization. The mechanism is synchrotron radiation, already introduced in Chapter 3.

Two Swedish physicists, Alfven and Herlofson, first proposed that some of the discrete sources of celestial radio emission might be stars with large magnetic fields, in which electrons with very high energies might be moving in curved tracks and radaiting synchrotron radiation. The idea that the sources were indeed stars was soon dropped, to be revived much later for some most interesting individual sources (see page 91), but the idea of synchrotron radiation was taken up by Soviet astrophysicists, and applied to the general interstellar radiation from the Milky Way. The theory was developed particularly by I. S. Shklovsky, who was able to show that both the light and the radio waves from the Crab nebula must be synchrotron radiation: his colleague V. L. Ginzburg then predicted that the light should on this account be fairly highly polarized. Experimental confirmation of this prediction came first from V. A. Dombrovsky and from M. A. Vashakidze in the Soviet Union. Here at last was proof of the only plausible hypothesis, and at the same time a challenge to radio astronomers to find a similar polarization in the Crab's radio emission. The result can be seen in Figure 15, which shows the latest map of the radio polarization obtained by R. G. Conway at NRAO (the National Radio Astronomy Observatories in the USA). The comparison and analysis of these optical and radio pictures should provide a detailed model of the magnetic field and the flow of electrons within the nebula.

The analysis of the optical emission raised a serious problem about the energy supply of the Crab nebula. There could be no doubt of the violence of the supernova explosion, but the electrons should have lost their initial energy, radiating it away as the light by which the nebula can be seen. Matters become worse when the X-ray emission was discovered, since this required the electrons to have such high energies that they would be expected to use up their energy within a few years only. There must be a hidden powerhouse within the nebula, stoking up the fires of the radiation continually – possibly a nuclear powerhouse providing a steady energy supply, in contrast to the nuclear explosion that started the nebula on its expansion. The solution to this problem was provided accidentally by the discovery of the pulsars, described in the next chapter. The most famous of these is in the centre of the Crab nebula, and it is known to be the source of the enormous energy which produces the light and the X-ray emission from the nebula. This pulsar is the only one which itself

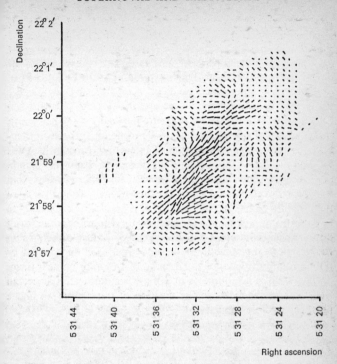

15. Polarized radio waves from the Crab nebula. The lines represent the strength and the orientation of the plane-polarized radiation. These lines reveal a high degree of organization in the magnetic field of the nebula (after R. G. Conway)

emits light and X-rays as well as radio waves, and it is the only pulsar which is clearly within a supernova remnant, so that the Crab nebula has again shown itself as a unique object. In 1970 it was the sole subject of a three-day conference of the International Astronomical Union, and it has been the sole subject of several books. It is not such a wild exaggeration to state that astrophysics seems to be divided equally between the Crab nebula and the other celestial objects.

Other Supernovae

The dates at which supernovae have been observed in the galaxy are the years A.D. 1054, 1572, and 1604. This series, it must be admitted, provides no reliable guide as to the frequency of occurrence of supernovae, as it has been followed by a blank period of 350 years. Supernovae can, however, be seen in other galaxies, and a fair guess at their frequency of occurrence in a spiral galaxy of our type indicates that one might be seen every few hundred years on the average. There will be others which will explode in our own galaxy without being seen from this planet. A supernova occurring in the galactic plane, unless it were very near to us, would be hidden by the dust which is concentrated in the plane. In fact, for every one seen, it is possible that as many as ten may occur unseen, and out of these there must be some which now are radio emitters. To search for these we must look for radio sources near the galactic plane, and then look for optical or photographic evidence of any remains of an explosion near them. Among such radio sources is one of the brightest radio objects in the sky, Cassiopeia A; in the position of this source we find a most remarkable nebula.

The Radio Nebula Cassiopeia A

In 1948 at Cambridge, an interferometer was built for the study of the Cygnus A 'star', already reported by Hey and by Bolton. The first record obtained showed that Cygnus A was outshone by another radio 'star', which was too far north in the sky to be visible to Bolton in Australia and out of reach of Hey's aerial which looked only at the horizon. A rough position was soon found, but no obvious identification was possible.

As a research student in the Cambridge team the author was given the task of finding an accurate position of this new radio star. The main objective was to try to identify this important source of radio waves with a visible star or nebula. A special interferometer was built, using two 27-ft parabolic reflectors from old German radar sets, and was operated at a frequency of 214 MHz. During a long series of observations, three other radio sources were also measured, among them the Crab nebula, whose identification was already fairly certain. The position of Cassiopeia A was eventually found, to an accuracy of 10 seconds of arc in Right Ascension, and 40 seconds in Declination (these are the positional coordinates for celestial objects, corresponding to longitude and latitude for terrestrial positions), and the position was passed on to the Cambridge observatories. With the optical instruments available Dr David Dewhirst was able to show that there was a very faint nebula near to the radio position, and Baade and Minkowski at Mt Palomar were therefore asked to take the optical comparison to its limits by checking the position with the 200-in telescope. The photograph of Plate X is the result of their cooperation in this piece of detective work.

One might expect the brightest radio star to be an outstanding visible object, but Cassiopeia A is certainly not this. In fact it is possible to miss it entirely at a first glance at the photograph, until one is given the clue that the nebula occupies almost the whole of the Plate. Near to the top of the photograph is a large blob of nebular gas, which is in fact the piece found by Dewhirst, and over an area of about 5 minutes of arc in diameter, there are more clouds and wisps of gas that are in fact all that can be seen of a sphere of hot gas.

In spite of the faintness of the nebula, Dr Minkowski managed to obtain spectra of several of the filaments of gas, and found that the nebula had some optical characteristics which placed it quite out of the ordinary, and indeed matched its radio performance. He found two kinds of filaments, one nearly stationary, and the other moving at speeds of around 5 000 km sec^{-1} (about 200 000 miles per minute). These move so fast that their detailed appearance changes appreciably in only a few months, and photographs at intervals of two years show some distorted almost out of recognition. No other massive body moves so fast in the whole of our galaxy. Further, the gas in the filaments is very hot. There are oxygen and neon atoms there which have been stripped of four electrons by the thermal agitation. A highly energetic, ionized, fast-moving gas is just the place for the generation of radio waves, and we can be fairly sure that there is a magnetic field as well, giving the conditions necessary for synchrotron radiation.

What then is this nebula? The first attempt to make sense of the motions of the filaments and to show that they were all streaming out from one place, as they would in a supernova, failed in this, and showed only a complicated, almost turbulent action. There was a spirited argument for a while, until, after some years, the motions could be followed more clearly from a succession of photographs. Then it became obvious that the fast-moving filaments were all flowing out fairly regularly after all, but that the others were part of a more chaotic system. The slow filaments also had a quite different spectrum, and it seems that they may be clouds formed when fast filaments collide with cold interstellar matter, and fall behind the rest of the explosion shell.

The 1-mile radio telescope at Cambridge has been used to give a detailed radio map of Cassiopeia A, shown in Plate X. A fairly symmetrical shell can be seen, with some irregularities which do not appear to coincide with the visible nebulae. The bright radio shell presumably is a region where a shock front of electrons is meeting a compressed interstellar magnetic field.

From the speed of the movements of the fragments, the explosion of Cassiopeia A as a supernova could be dated to about A.D. 1702, within a few years. Its distance is 10 000 light-years, and this, placing it well away from our own position in the galaxy, could explain why it was not seen and recorded as a supernova along with those of A.D. 1054, 1572, and 1604. It appears to have been a rather different and more powerful type of supernova explosion than the other three, but at a distance of 10 000 light-years it might only have reached 5th magnitude, which is only just visible to the naked eye.

There must, of course, be others like it, although no other in our galaxy is quite as powerful. Cassiopeia A, placed at the most distant part of the galactic plane, would still be a very prominent radio source, brighter than any others that have been found in the galaxy. The supernova observed by Tycho Brahe in 1572 has left a very clear radio remnant, seen in Figure 16. This is a weak radio source compared with Cassiopeia A, but it shows as a beautifully symmetrical expanding shell.

Another identified radio source is the nebula known as IC 443, from its position in the Index Catalogue of nebulae (Plate XII).

This beautiful object is much less energetic, both as a radio transmitter and in optical appearance; nevertheless it is hard to believe
that such a well-defined and symmetrical cloud could have come
from anything other than a supernova. It may be that the explosion was less violent or that it occurred in a region of high density
gas which has blanketed and slowed down the expanding shell.

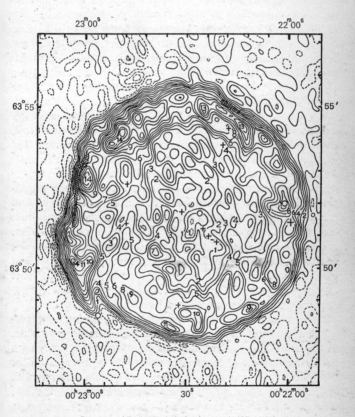

Remnant of Tycho Brahe's Supernova (1407 MHz)

16. Radio remnant of Tycho Brahe's supernova. The expanding shell is
seen only in its radio emission (Cambridge 1-mile telescope)

More likely it is very much older than Cassiopeia A, and has spent most of its energy. Radio waves do not come uniformly from the whole of this nebula. There is a bright filament near the centre, which is an especially powerful patch; perhaps this has not yet lost so much of the explosive force of the supernova.

It is possible that other less obvious radio sources are also parts of supernova outbursts. The 'Belt', or 'North Galactic Spur' (see page 110) is probably part of a rather diffuse shell seen quite close by; other blobs and streaks of radio emission near the galactic plane may also be parts of such nebulae.

We cannot tell how many of these remains there are in our galaxy, but they are certainly not uncommon, and some of them are certainly young. The concept of an unchanging universe implies little change over periods of at least a million years, but in our galaxy we apparently have nebulae as young as a few hundreds of years old, and another could be born tomorrow, without warning. Nothing could please astronomers more! The heavens as they are are exciting enough, but the birth of a new supernova in our galaxy would be headline news all over the world. Every telescope would be trained upon it, and photographic plates and films used at a rate at least as great as that expended on a new Hollywood star. Radio astronomers would compete to be the first to detect any build-up of radio emission, and then would follow its rise as the hot sphere of gas grows, generating its own magnetic field, and giving a spectrum of radio waves which, one hopes, would reveal some of the complex dynamics of such an energetic cloud.

The stir created by the Russian launching of Sputnik I would seem trivial in the face of such excitement, and no doubt journalists and astronomers alike would cheerfully lose sleep, as then, in the cause of its observation. But, however much we may wish for a supernova to appear in our own lifetime, these things are outside our control and for the moment we content ourselves with watching Cassiopeia A. The visible parts of this are moving fast; does the radio source also change?

There have been some good measurements of the apparent size of the radio source, as for example in the map of Plate X, but there has not yet been time for the expansion to be clearly demonstrated. The total intensity of the radio emission seems to be falling by about $\frac{1}{2}$ per cent each year; a repetition of the diameter measurements should soon show whether the radio and light sources are expanding together.

Flare Stars and Radio Stars

The first glimpses of the radio sky by Hey and his early followers suggested that there were point sources of radio waves, which by obvious analogy were then called 'radio stars'. The first four identifications of radio sources turned out quite differently; two were supernova remnants (the Crab nebula, and Cassiopeia A), one was a conspicuous extragalactic nebula (Centaurus A) and the fourth was a very distant and inconspicuous galaxy (Cygnus A). The idea of a true radio star was quietly dropped in favour of names like 'radio nebulae' and 'radio galaxies'. Even when some remarkably star-like objects were identified as radio sources, these proved to be the bright centres of galaxies, and only stellar in appearance because of their great distance. Recently, however, some genuine radio stars have been discovered.

Sir Bernard Lovell was the first observer to follow a suggestion by E. Schatzman, of Paris, that there might be an analogy between flares on the sun and the much larger variations of light given out by a particular class of red-dwarf stars, so that a visible flare on the star might be accompanied by a radio outburst. After many hours of patient observing in combined sessions with the Mark I radio telescope at Jodrell Bank and optical telescopes at the Smithsonian Institute in the USA and at the Crimean Observatory in the USSR it was established that radio flares could be detected. The radio signal is, however, too small for much detailed study to be made of it.

More recently several new types of radio star have been detected by C. M. Wade and R. M. Hjellming, using a sensitive radio interferometer at Green Bank, USA. In their first observations they followed the idea of the flare stars by looking at some other exciting visible stars, and particularly at two novae. These are stars which become suddenly very bright but without the complete explosion of the whole star, which then becomes a supernova. Only about 10^{-5} of the mass is lost in the flare-up of a nova which lasts about a year. These two are known by their constellations and dates as Nova Delphini 1967 and Nova Serpentis 1970. Both of these were found to be detectable as radio sources with the NRAO interferometer, and both are variable in strength.

Encouraged by this success the same observers have started to compile an interesting list of radio stars. The bright stars Algol and Antares are amongst them. But evidently a star must have some uncommon qualities to be a radio emitter, since most of the familiar stars are undetectable. A good clue came from accurate positional measurements on the giant star Antares, when it appeared that the radio waves came not from the bright star itself, but from a very insignificant companion, a blue dwarf rotating about it. Pairs of stars, known as binaries, are very common, but this particular combination of star types, with a very small distance between them, is not at all common. Furthermore, the massive dwarf star is believed to be extracting gas from the giant by gravitational force. In this binary, therefore, there is the usual requirement for non-thermal radio emission – hot, ionized, streaming gas. Here again is a celestial laboratory containing conditions which we cannot produce on earth, with the promise of interesting advances in plasma physics. (A 'plasma' in physics is an ionized, neutral gas, which is regarded as a kind of hotbed on which all sorts of processes, oscillations and instabilities can grow.)

The next stage in this story of the radio stars has been achieved through X-ray astronomy. Observing from X-ray detectors in satellites such as the USA satellite UHURU (Swahili for freedom, commemorating its launch from the coast of Kenya) has shown that there are some hundreds of 'point' sources of X-rays in the sky. These are being identified as soon as they have sufficiently accurate positions, and about a quarter of them turn out to be stars which are similar in many ways to the radio stars. Furthermore, some of the X-ray sources are pulsating, with periods of a few seconds; several, and in particular one known as Hercules X-1, also vary with a period of a few hours. We shall see in the next chapter that the rapid pulsation is typical of a rapidly rotating and very condensed star known as a neutron star; the classic example is the pulsar in the Crab nebula, which emits pulses of both X-rays and radio. Unfortunately no other pulsating X-ray source has been observed as a radio pulsar, but there can be no doubt that they all involve neutron stars.

In contrast to the radio pulsars, the X-ray sources are usually binary systems, in which a neutron star is orbiting round a larger normal star. The orbital period is a few hours, as observed in the variable emission from Hercules X-1. The neutron star is very close to the surface of the normal star, and it can pull gas out of the star by gravitational attraction. As this gas falls onto the neutron star it becomes hot, so that X-rays are emitted. This thermal radiation is usually too weak at radio wavelengths to be seen as a radio source. Only weak and randomly fluctuating radio emission has been discovered from any X-ray source apart from the Crab nebula pulsar and another remarkable exception known as Cygnus X-3.

In August 1972 the radio emission from Cygnus X-3 suddenly increased by a factor of about one thousand, making it the brightest radio nova ever observed. The growth and subsequent decay of the radio emission were observed by astronomers all over the world in a remarkable cooperative effort (see Chapter 17). The origin of the outburst is unknown, but the result was an expanding cloud of plasma, radiating synchrotron radiation and following a classic pattern predicted for a supernova explosion. This was not, however, a full supernova, as the X-ray source is still there almost unchanged. Cygnus X-3 may be capable of more outbursts of this kind, and it will be watched closely over the next few years.

CHAPTER 7

Pulsars

THE announcement in February 1968 that pulses of radio waves, as regular and as rapid as a ticking clock, were being received from the Milky Way, astounded the whole astronomical world. There was no periodic phenomenon known to optical astronomy which could match the rapid repetition of the pulses, which occurred at intervals of about 1 second. Only a small object, some sort of very condensed star, could rotate or vibrate with such a short time-scale; none such had ever been observed, and the only experience of such an accurate periodicity referred to much larger and slower systems such as planets or binary stars in orbits with periods of days or years. Furthermore, no mechanism of radio emission was known which could produce intense broad-band pulses only a few milliseconds long. It is small wonder that the first appearance of periodic radio pulses on the records of a new radio telescope at Cambridge was greeted with scepticism, which grew into astonishment as their reality and their remarkable properties became inescapable.

Most celestial sources of radio waves produce a steady signal, and the radio receivers normally used for their detection accordingly average the signal over several seconds to obtain a good accuracy of measurement. Pulsars cannot, however, be detected in this way, and it was only when a receiver which was deliberately given a rapid response time was used with a large radio telescope that the very short radio bursts from the pulsars were discovered. The receiver, and the telescope, were specifically designed for the study of interplanetary scintillation (see page 65).

Professor Anthony Hewish has for several years carried out in Cambridge a systematic investigation of interplanetary scintillation. In the summer of 1967 he completed the construction of a new radio telescope designed for the repeated observation of a large number of quasars throughout the year, as the sun and its corona moved across the celestial sphere. This new telescope was made by covering $4\frac{1}{2}$ acres (2×10^4 m^2) of ground with a network of dipoles, all connected together to a single receiver. It was designed for a wavelength of 3·7 m, which is rather long in the range of wavelengths used in radio astronomy. This choice was made so that the recordings would show clearly the effects of interstellar scintillation, and particularly how this scintillation might vary through the year. It was this telescope which gave the first recordings of the pulsars. In fact it proved to be admirably suited to the detection of the pulsar signals, which are most easily detected at long wavelengths and which require a receiver which is sensitive to rapid fluctuations.

The strange pulsating signals were first noticed by Miss Jocelyn Bell, a research student who was helping Hewish by scanning the records of the scintillating quasars. She found the record of a new source, apparently varying very rapidly, but in a direction away from the sun where not much scintillation was expected. The strength of the signal was different in each day's recording, and there were many days when nothing could be seen. Hewish at first suggested that the recording merely represented some form of radio interference, perhaps from a passing motor car, but Miss Bell was insistent that the source had a fixed position in the sky, and that it was a new type of celestial object. Eventually a specially fast recording was made, and the rapid fluctuations emerged as the periodic pulses shown in Figure 17.

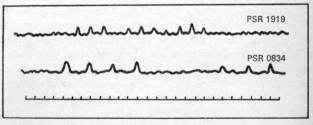

17. Radio pulses from two of the pulsars discovered at Cambridge. These early recordings are labelled with the designation PSR, followed by the position of the pulsar in the sky. The time scale is marked in seconds

Even at this revelation there was some scepticism and a good deal of alarm. Could the pulses come from a spacecraft? or even from an extraterrestrial civilization? Only a few years previously a team of radio astronomers in America had spent many observing hours trying to pick up radio signals characteristic of life on other worlds – the 'little green men' of science fiction. Here perhaps was the first success, or sufficiently near to it to bring an uncomfortable and unwelcome deluge of publicity to the observatory. So Hewish and Miss Bell kept quietly on until three other pulsars had been discovered, and until some basic characteristics of their radiation had told them that they were stars, not people. Nevertheless, the thought haunted them for several months; they even gave expression to it in their first catalogue numbers for the four pulsars, known as LGM 1, 2, 3, and 4. Today the catalogues list them as PSR; pulsating sources of radio, rather than little green men.

The true nature of the pulsars is scarcely stranger than the hypothesis that they were the product of another civilization. They are now known to be neutron stars, the end point of stellar evolution and the strangest objects in the Milky Way.

Neutron Stars

The smallest stars which had been observed optically at this time were the white dwarfs. These are old stars in which energy generation has almost ceased, so that outward radiation pressure from the interior could no longer prevent collapse of the star by its own internal gravitational forces. The collapse of a star like the sun to a white dwarf is a shrinkage by a factor of over 100, condensing the whole mass down to a sphere about the size of the earth. White dwarf stars are so dense that a matchboxful would weigh about a ton.

A further stage of collapse had already been predicted by the Soviet scientist L. Landau in 1932. In a white dwarf the atomic nuclei are tightly packed together, but in the next stage, known as a neutron star, the nuclei themselves collapse and the basic particle becomes the neutron; when these are tightly packed a further shrinkage takes place so that the whole mass is only 20 km across. We might compare this to the size of Mt Everest rather than the whole earth. The density is now such that a pin head of neutron star material would weigh as much as an ocean liner.

The trouble with the theory of neutron stars was that it suggested that they could never actually be seen, as they are too small. How then can they ever be identified with the pulsars?

The main clue comes from the very short time-scale of the radio pulses, and from the clock-like precision of their timing. The accuracy of the period might be determined by the vibration of a star, or by its rotation. But a star as big as the sun, or even a white dwarf, cannot vibrate as fast as once per second; neither can it rotate as fast as once per second, or it would burst apart by centrifugal force. A neutron star, however, could rotate up to hundreds of times per second without bursting; indeed it would be expected to be rotating very rapidly as a result of its collapse from a larger and slowly rotating star like the sun. The 'clock' rate could easily cover the known pulsation rates of pulsars, which range from 30 per second to 1 per 4 seconds. There is no other conceivable celestial body that could do this.

With the revelation that the pulsars must be the long-dreamed-of neutron stars, there appeared a remarkable link with the Crab nebula, described in the previous chapter. This nebula was known to contain a source of energy, strong enough to provide power for the continued emission of light, X-rays and gamma-rays 900 years after the supernova explosion had produced the expanding nebulous cloud of gas. An Italian theorist, Franco Pacini, had already pointed out that a rapidly spinning neutron star contained sufficient energy for the missing power-house, and that it could drive the electrons in the nebula to high energies if, as seemed likely, the star had a strong magnetic field. It would then act like an electrical dynamo, a kind of celestial turbo-alternator. As it gave up energy to the nebula, it would slow down, but it ought to have enough energy to last for some thousands of years at least.

This suggestion led to a search for pulsars in supernova remnants generally. Success came first to the Molonglo Observatory in Australia. A young nebula in the constellation of Vela, in the southern sky, was found to contain a pulsar with a pulse rate of just over 11 per second, the fastest yet known. Further, it was soon found that the rate was slowing down, according to Pacini's prediction (Figure 18). Then the Crab nebula pulsar was found, through a combination of observations at the Greenbank and Arecibo Radio Observatories. This pulsar gives 30 pulses per second: it is in fact a neutron star spinning at about the same speed as the dynamos which power the 60-cycle electricity supply in the USA. Again the rate of pulsation was found to be slowing

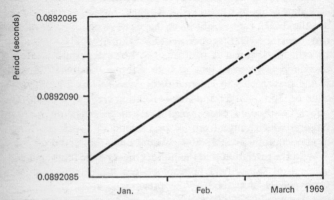

18. The changing period of the Vela pulsar. The period increases uniformly with time, apart from the step at some time between 24 February and 3 March, when the period decreased by 200 nanoseconds

down, this time at a rate consistent with the known age of the nebula. (The change in period is only 0·000000300 seconds per day, in a period of 0·033 seconds, but the effect is measurable even during an observation period of only one hour.)

Pulses of Light from the Crab Nebula

The discovery of the Crab nebula pulsar, the most exciting pulsar of them all, immediately led to speculation that this might be one of the visible stars at the centre of the nebula, and in particular one star which has a very unusual spectrum, showing a continuum of light with no spectral lines. This was already thought to be some form of remainder of the supernova explosion, although it seemed impossible that it could be a neutron star as it would then be far too small to be seen. Could the light from this star possibly be pulsed light, far brighter than ordinary starlight, flashing on and off 30 times every second? This was looked for, and quickly found to be true, by observers at the Steward Observatory in Arizona. At last a real galactic lighthouse had been discovered.

The photograph of this astonishing star in Plate IX was taken at Lick Observatory by superposing very short exposures recorded by a television technique. Two parts of the pulsar cycle were recorded separately, one on the flash and the other a few milli seconds later. There is a ratio of at least 100 between the brightness of the star on and off the light pulse.

If it radiates light, why not X-rays as well? The experimental test here needs a rocket to carry the apparatus above the atmosphere; despite this difficulty two independent experiments were carried out in the USA within weeks of the discovery of the light pulses, and strong X-ray pulses were found. Furthermore, some recordings made two years previously in a balloon-borne experiment were re-analysed, and what was thought at that time to be continuous X-ray emission from the Crab nebula was found to contain a component pulsating 30 times per second which could only have come from the pulsar.

The Glitches

The Vela pulsar, PSR 0833, was the first to show the steady increase in period which proved that pulsars are rotating neutron stars. It was also the first to show a departure from a smooth increase of period. Figure 18 shows that observations taken only a few days apart in February 1968 showed that the pulsar had speeded up by two parts in a million, and then resumed its steady slowing. This unexpected jump was called 'the glitch'.

Time is the most accurately measurable of the basic physical quantities, so that such a small step is easily found. The Crab nebula pulsar shows rather better timekeeping than PSR 0833, but it was soon found to be considerably worse than our atomic clocks. Apart from the smooth change in period, there are irregularities over days and months, in which the pulses arrive late or early by up to 150 microseconds. Worse still, there was a glitch on 24 September 1969, when the error built rapidly up almost to one whole period of 30 milliseconds.

A rigid body cannot behave in this way. Any departure from a smooth slow-down means that the star's moment of inertia must have changed abruptly, which in turn means its shape has changed. The theory of neutron stars now allows for this in some extraordinary details of the behaviour of over-dense matter. The inside of the star is now known to be a liquid, with a solid crust occupying a depth of about 1 km. The glitch is a sudden change of shape in this crust – a 'star-quake' – while the smaller and slower timing changes represent a variable slippage between the rotation of the liquid core and the solid crust.

How Do Pulsars Pulse?

Now that the pulsars are known to be rapidly spinning, highly magnetized, very dense stars, attention is turning to the nature of the radiation mechanism. We know that the radiation leaves as a searchlight beam, rotating with the star, and that this beam has very similar characteristics over the whole electromagnetic spectrum (Figure 19). There have been two schools of thought about the formation of the beam. Either it is formed within the radiation process itself, as for example a terrestrial radio transmission can be beamed, or it is formed by a previously unknown process, called 'relativistic beam compression'. Pulsars are such unusual objects that it is not surprising to find the evidence accumulating in favour of this new process.

Figure 20 shows a cross-section of the magnetic field surrounding a pulsar. The shape is probably over-simplified, but there will be some such field rotating with the star. It is a very strong field, about a million million gauss, and it also generates a very strong electric field as it rotates. This drags material out of the pulsar, forming a dense atmosphere of energetic charged particles extending far out from the pulsar. The atmosphere rotates with the pulsar, being swept round by the magnetic field as though

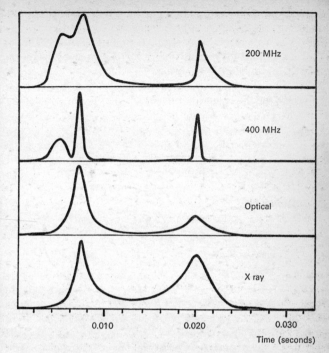

19. The pulsed emission from the Crab nebula pulsar at radio, optical and X-ray wavelengths. At 200 MHz the pulses are broadened by interstellar scintillation

the field lines were the spokes of a wheel. Somewhere in this co-rotating atmosphere, on some fixed point in the magnetic-field pattern, the ionized atmosphere is oscillating and acting as a radio transmitter. This transmitter is swept round as the pulsar rotates.

The further out from the surface of the star, the faster the atmosphere must move. If this co-rotation extended far enough outwards, the speed would reach the velocity of light, which is a fundamental impossibility. So there must be a breakdown somewhere inside the 'velocity-of-light circle' shown in Figure 20.

101

The transmitter is located in, or close to, this breakdown region; it is moving at about four fifths of the speed of light. Relativity theory tells us that this high speed has a most remarkable effect on the radiation, forming it into a searchlight beam pointing ahead in the direction of motion. This relativistic beam compression applies equally to radio, light, X-rays and gamma-rays, accounting for the similarity of the pulses in Figure 19, despite

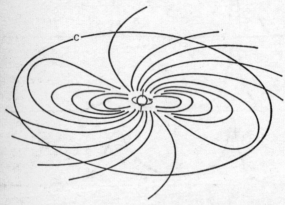

20. Magnetic field round a rotating neutron star. This idealized sketch shows the lines of force from a dipole swept back by the rotation. At the 'velocity-of-light circle' c the pattern moves with velocity c; inside this circle there is an ionized atmosphere rotating with the star

the fact that they are observed over a frequency range of no less than 39 octaves.

This still leaves open the nature of the radiation process itself. Having found the size and location of the transmitting region, it is a simple matter to calculate its total power and the 'volume emissivity', or the rate at which power is radiated per unit volume. Light, X-rays, and gamma-rays account for most of this power in the Crab nebula pulsar. The radiation process for this range of the spectrum seems to be the familiar synchrotron radiation; for radio there is an added complication of some coherent motion of electrons as already seen in the solar corona.

But the remarkable conclusion is that the radiation object is so powerful and so small that it has a higher energy-density inside it than has the core of a nuclear reactor.

Characteristics of the Pulses

The radio pulses emitted by the pulsars are by no means simple bursts of radiation repeating precisely at each rotation. If some hundreds of pulses from a pulsar are superposed, a characteristic shape emerges. As shown in Figure 21, this 'integrated pulse envelope' may for different pulsars be single, double, or complex. Some pulsars, notably the Crab nebula pulsar, show two pulse components per revolution. Individual radio pulses, by contrast, are very variable in shape, size and time of arrival. They are also very highly polarized, and for most pulsars the polarization is very variable. There is here a complicated story to unravel when the radiation mechanism is better understood.

On top of these variations, which are occurring within the radio emitter itself, there are some very important effects which occur after the radio pulses have left the pulsar. As they travel through interstellar space, the radio pulses traverse ionized gas, consisting mainly of protons and electrons. This gas transmits radio pulses more slowly than does free space, with the result that a radio pulse may arrive typically some minutes later than it would if it travelled in free space. Further, the delay depends on the radio frequency at which the pulse is observed, so that simultaneous observations at two frequencies will show different delays, and the total delay is easily obtained from the difference. The delay depends on the total number of electrons along the line of sight between the pulsar and the observer; if the electron density is known this therefore gives the distance of the pulsar.

Further effects of the interstellar medium, namely scintillation and Faraday rotation, will be dealt with in Chapter 8.

The New Physics of Pulsars

The chance discovery of a new type of fluctuating radio signal in Hewish's scintillation experiment has opened up a new physical

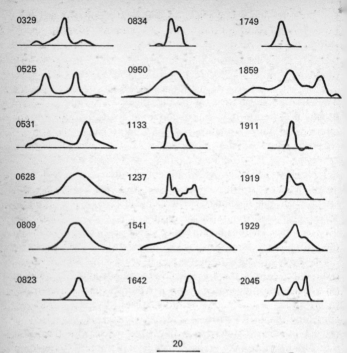

21. Integrated pulse shapes from different pulsars, obtained by adding some hundreds of individual pulses from each pulsar. The time scale is calibrated in degrees of rotation, so that the profiles represent cross-sections of the 'lighthouse' beam of the rotating pulsar. The PSR catalogue numbers of the pulsars indicate their positions in the sky

world, with extreme conditions quite unattainable in terrestrial laboratories. At the surface of a pulsar there is a gravitational force more than 10^{10} times that on earth, and there is a magnetic field more than 10^{12} times that on earth. Within the pulsar there is matter with density at least 10^{14} times that of terrestrial matter. Outside the pulsar there is a radio transmitter generating a power of some megawatts per cubic metre, and a large enough energy flow to keep a whole nebula glowing for thousands of years. This is a field for physics of a new kind, not perhaps with any new fundamental principles, but so far outside our previous experience as to provide new illumination in familiar problems. No one can tell what this new illumination will bring. It may be that our first glimpse of matter at neutron densities will bring discoveries in the physics of fundamental particles. More likely, the new plasma physics of the pulsar magnetosphere will eventually find its way into the design of fusion reactors, which must eventually provide the major source of power on our own planet.

Astrophysical Uses of Pulsars

The pulsars are not only exciting for their own properties and strange physical conditions. The sharp, highly polarized radio pulses from this new form of celestial radio beacon have been found to be the most penetrating and informative probes of the interstellar medium. In the next chapter, it is shown that the radio pulses can give entirely new information on the ionized interstellar gas and on the magnetic field that pervades it. But it should be noted here that these studies have themselves begun to react back on the understanding of the pulsars. Scintillation studies have now developed into a study of the velocity of pulsars, which has shown that at least some of them are travelling through the interstellar gas at speeds of several hundred kilometres per second, faster than almost all known stars.

The most likely explanation of these high velocities is that the pulsars are indeed generated in supernova explosions, and that the explosion is sufficiently asymmetric to shoot the pulsar away like a rocket. Again, this is a new piece of astrophysics which still remains to be explored.

How Many Pulsars in the Milky Way?

The distances of pulsars can be measured fairly well by the magnitude of the pulse-delay times, and for some pulsars quite accurately by a study of the absorption of their radiation in the neutral hydrogen contained in the Milky Way (Chapter 8). A comprehensive search for pulsars over the whole sky should therefore tell us how they are distributed through our galaxy. Furthermore, we can measure their ages from the rate at which their periods are lengthening, so that their place of origin and their history ought to be readily available.

First attempts at this piece of cosmogony suggest that a new pulsar begins life in our galaxy approximately every hundred years. This is remarkably close to the rate of supernova explosions, which lends support to the idea that the Crab nebula pulsar is the typical pulsar, born at the time of the supernova explosion, and seen in the early years after its formation. If pulsars live typically for 10 million years, and one is born every 100 years, there should be about 100 000 pulsars in the galaxy. Of course, most of these will be too far away to be detectable, but it does seem that a good many could be found.

Complete surveys of the sky are difficult. An observing method adopted at Jodrell Bank is described in Chapter 16, where it will be seen that to search the sky with sufficient sensitivity, and covering a sufficient range of periodicities, is complicated and time-consuming. At the end of 1973 the total of known pulsars has reached 105, most of which were found in carefully calibrated surveys at Molonglo for the southern hemisphere and at Jodrell Bank for the northern hemisphere. These pulsars are concentrated strongly towards the plane of the Milky Way, and also towards its more central regions. So far, this evidence agrees with the hypothesis that pulsars are all born in supernova explosions. The troubles arise when the more direct test is made, by looking for pulsars within the known and obvious supernova remnants, of which there are at least a hundred. Only two, the Vela and the Crab remnants, are found to contain detectable pulsars. Where are the others?

This mystery was solved in the study of pulsar scintillation, which showed that pulsars move with very high velocities. The experiment was international, carried out by A. Lyne at Jodrell Bank and J. Galt at Penticton, Canada. The large spacing between these observatories enabled them to show that the scintillation pattern was moving very rapidly over the earth, giving a velocity of 370 kilometres per second (km sec^{-1}) for the pulsar PSR 0329+54. At such a high velocity a pulsar would move out of its parental supernova remnant after only 10 000 years or so, and only the very young pulsars would still be associated with the supernovae where they were born.

Confirmation of the idea of runaway pulsars was soon obtained. The nebula IC 443 (Plate XII) is a supernova remnant about 60 000 years old. In one of the Jodrell Bank surveys a pulsar was found just outside the nebula, and its age was found to be the same as that of the nebula. This is an unusually small age for a pulsar, so that it must be associated with the nebula. To have reached its present location it must have been moving at over 100 km sec^{-1}, which agrees well with the velocities determined for other pulsars from their scintillations.

CHAPTER 8

Interstellar Gas

IF from the best vacuum we can obtain, we take away all but one in a thousand million of the remaining particles, we have something resembling the emptiness of interstellar space. It is small wonder therefore that the gas between the stars is hard to detect optically. It is dispersed in a far better vacuum than any we can create on earth. Nevertheless the material it *does* contain constitutes a large proportion of the total mass of the Milky Way; furthermore it radiates radio waves in several different ways, each of which can be used for studies of the whole galaxy as well as for revealing the conditions in interstellar space.

Synchrotron Radiation from the Milky Way

The major part of the radio waves received from the sky is emitted by electrons with very high energies moving freely throughout the Milky Way. These are the cosmic-ray electrons, which radiate synchrotron radiation as they encounter the magnetic field of the galaxy (Chapter 3). This is the radiation which was discovered by Jansky in 1932. Modern radio telescopes have made very much more detailed maps of this radiation; the whole sky has been covered by combining observations from observatories in the northern and southern hemispheres, giving us the map of Figure 22. This map is displayed as a kind of opened-out sphere, whose equator runs along the bright circle of the Milky Way. The top and bottom of the map represent the poles of the galaxy, perpendicular to the Milky Way itself.

Maps of this kind have been made for several different wavelengths through the radio spectrum. The main problem in drawing the maps with sufficient accuracy is the limited angular resolution of most radio telescopes. Most of the galactic radiation is emitted at long wavelengths, 1 m or more, where the beamwidth of even the large radio telescopes such as the Jodrell Bank 250-ft

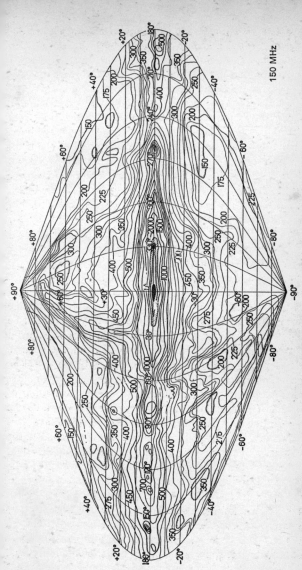

22. Galactic radio map at 150 MHz. The contour map is plotted on galactic coordinates, with the Milky Way along the equator (from the work of Landecker and Wielebinski)

telescope is about 1°. The map in Figure 22 has a resolution of 2·2°; the effect is that the map appears to be smoothed over, so that details smaller than 2·2° across are washed out. We now know, however, that most of the features appearing on the contour map of Figure 22 are shown in sufficient detail, since they represent large objects which do not contain much fine detail.

The map shows immediately that most of the radio waves come from our own galaxy. The line of the Milky Way shows prominently as a feature running along the centre of the map, exactly along the line of the bright stars. We see, however, an unfamiliar concentration to a peak at the centre of the map, which is the direction of the centre of the galaxy (see Figure 2). There is in fact a strong concentration of stars here also, but they are kept from the view of our optical telescopes by the opaque dust in the plane of the galaxy (Plate VI).

We do not expect to see spiral arms in this map, as we are seeing the galaxy 'edge on'. Furthermore, we are within the galaxy, so that we expect to see radiation from all round us. Most of the radiation shown near the poles of the map seems to be generated quite locally in the galaxy, probably within the spiral arm which contains the sun; unfortunately we cannot be sure about this, and there could be a much more widespread 'halo' of radio emission spreading far out from the galactic plane.

There are other features on the maps which need explanation in terms of something rather different from the distribution of visible stars. The most obvious of these runs up towards the north pole of the map from near the centre of the plane. It shows up in more detail on the map of Figure 23. It is often called the North Polar or Galactic Spur, or Belt, and it appears at first sight to be a branch of the Milky Way reaching up towards the pole of the galaxy. It is, however, better described as part of a circular arc or loop. There are other less conspicuous loops, tracing out fairly accurate circles in the sky. The North Polar Spur actually has a sharp outer edge, which in the map is somewhat blurred even by the 1° telescope beam of Figure 23.

Most probably the spurs and loops of radiation are fairly local features in the galaxy, which are conspicuous only because they

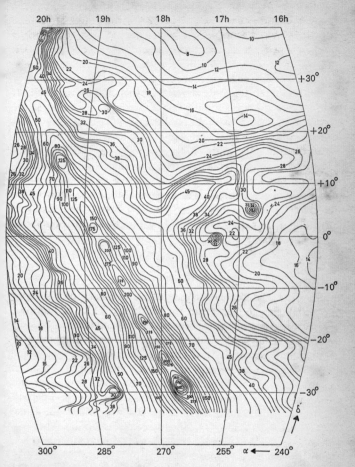

23. Radio emission from the Milky Way, plotted on celestial coordinates. Contour units indicate radio-brightness temperature. The ridge of the Milky Way runs from the galactic centre at the bottom of the map, towards the top left. The North Galactic Spur projects to the right (part of a survey at 70-cm wavelength, Dwingeloo, Netherlands)

are so close at hand. The circular form, with sharp outer edge and a more irregular interior, resembles the young supernova remnants like Cassiopeia A (Chapter 6); the main difference is that they appear to be very much larger. The present view is that they are expanding shells of gas, with electrons and a magnetic field included, which are blowing out from the remains of a supernova explosion only a few hundred light-years away. If this is correct, there should be an appreciable expansion in the North Polar Spur within the lifetime of present-day radio astronomers.

All these bright features radiate strongly over the long-wave-length part of the radio spectrum, through the synchrotron-radiation mechanism. There are other types of radiation which give us more details of some local parts of the Milky Way where there is hot ionized hydrogen gas, and about the spiral-arm structure of the galaxy, which is shown up by the distribution of cold hydrogen.

Hot Hydrogen – the H II regions

Most of the features on the map of the radio sky bear little or no resemblance to features of the visible sky. There is, however, a very close correspondence between optical and radio astronomy in one particular kind of interstellar cloud. Surrounding many of the brightest stars is a hot cloud of ionized hydrogen which appears on both radio maps and optical photographs. For most celestial objects the information available through optical and radio astronomy is complementary and often entirely different, but for the H II regions the information is very much the same, so we need not spend much time on their description. They do, however, provide the classic example of thermal radiation (page 29), which enables radio astronomers to measure the temperatures and densities of the clouds in a rather direct way.

These clouds are familiar to optical spectroscopists, who deduce from their characteristic red colour that they contain ionized hydrogen, known as H II, as compared with the cooler, un-ionized hydrogen (H I) which fills the major part of inter-stellar space.

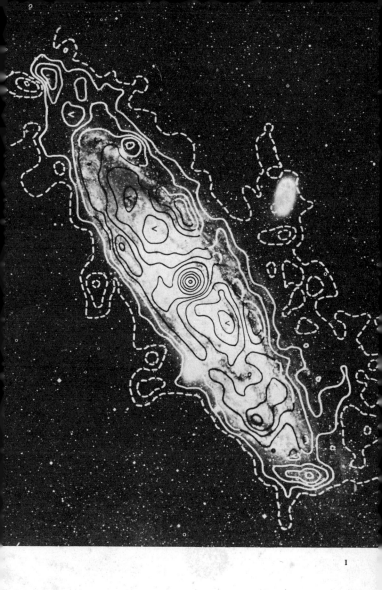

I   The Andromeda nebula (M31). Contours of radio brightness
(Cambridge 1-mile telescope at 21-cm wavelength) are superposed on
the Palomar Schmidt photograph (Hale Observatories)

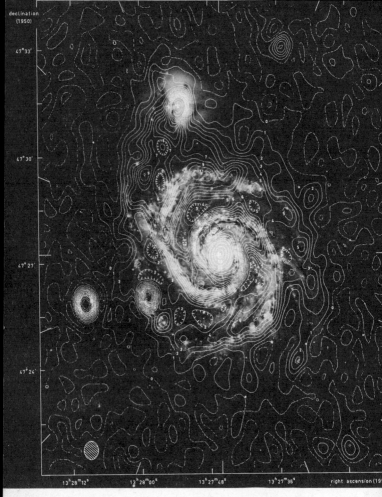

II The Whirlpool nebula (M51). Contours of radio brightness (Westerbork Array) are superposed on the Palomar Schmidt photograph (Hale Observatories)

III The Virgo A radio galaxy (M87). Peculiar elliptical galaxy in Virgo. This short-exposure photograph shows the nucleus of the galaxy, with the bright jet which emits polarized light and radio waves (Lick Observatory: RAS 608)

IV The quasar 3C 273. A negative photograph, greatly enlarged. The white dot in the centre represents the radio quasar component, and the white oval shows the jet, 19 arc seconds away from the centre (RAS 682)

V

VI

VII

V   A group of four extragalactic nebulae in Leo (Hale Observatories: PAL 19)

VI   The Milky Way. Photograph of the southern sky (Carina to Vulpecula) using a wide-angle camera (Schmidt-Kaler, Bochum Astronomical Institute)

VII   Orion nebula. A hot mass of ionized hydrogen gas in our own galaxy. This nebula is visible in the Sword of Orion (PAL 103)

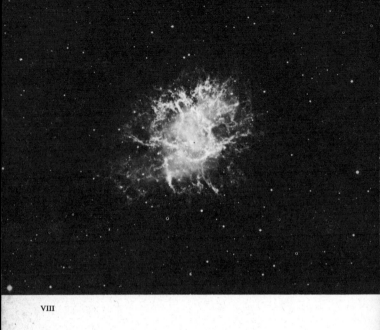

VIII

VIII   Crab nebula (M1, the Taurus A radio source). The remains of a
supernova explosion observed in A.D. 1054 (Hale Observatories:
PAL 7)

IX   The Crab nebula pulsar. Stroboscopic television photographs of
the pulsar at maximum and minimum phases of the 33-millisecond
pulse cycle (Lick Observatories: RAS 685)

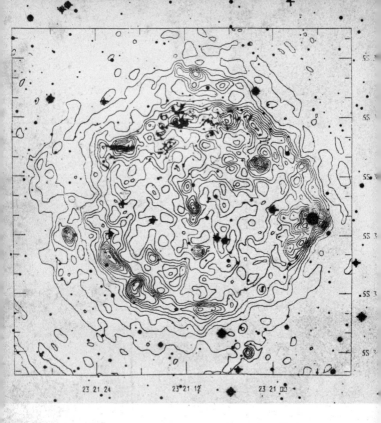

X   The supernova remnant Cassiopeia A. Radio contours superposed
on a photographic negative (Cambridge 1-mile telescope at 6-cm
wavelength)

XI   Cygnus A. A radio galaxy, the most powerful radio source known
(Hale Observatories: PAL 53)

XII   IC 443. A radio source, believed to be the remains of a
supernova explosion about 50 000 years old. A pulsar of the same age
is located just outside this nebula (Hale Observatories: PAL 56)

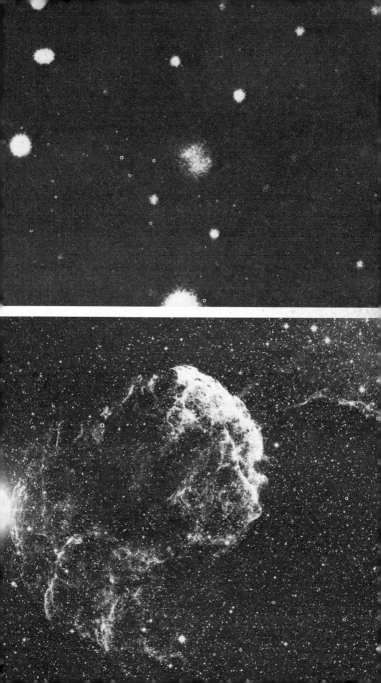

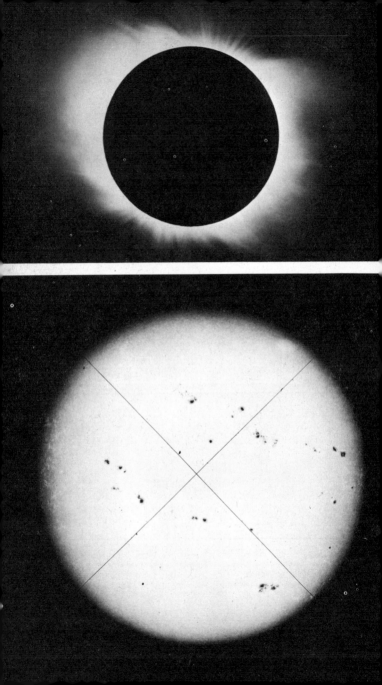

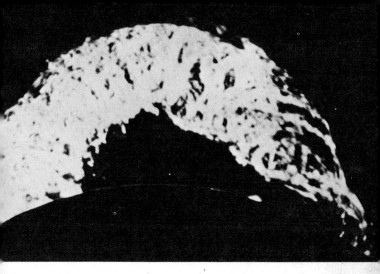

XIII   The solar corona. The corona is seen at a time of total eclipse (RAS 80)

XIV   Sunspots, 24 May 1947 (Royal Greenwich Observatory: RAS 548)

XV   Arch prominence, 4 June 1946. This eruption, following the curve of magnetic-field lines, resulted from a solar flare (Harvard-Climax Coronagraph: RAS 529)

XVI

XVI   Comet Bennett, 4 April 1969 (R. L. Waterfield's Observatory, Woolston: RAS 715)

XVII   Jupiter. Photograph in red light, showing the satellite Ganymede and the shadow it casts upon the planet (Hale Observatories: PAL 280)

XVIII   Meteor echo. Each vertical trace is a radar pulse echo, spaced at 20-millisecond intervals. The echo decays over $\frac{1}{2}$ second, showing slow oscillations of strength (Sheffield University)

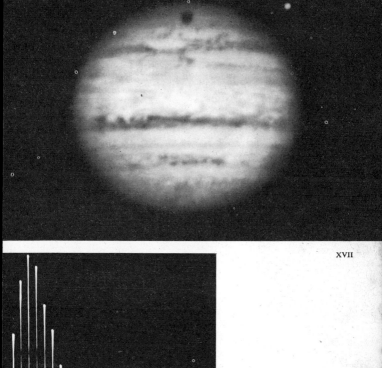

XVII

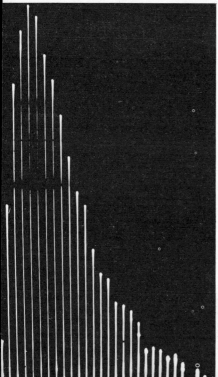

XVIII

XIX

XX

XXI

XIX   The Mills Cross radio telescope, showing the east – west arm.
The reflector surface is a cylindrical paraboloid

XX   Effelsberg 100-m reflector telescope, Max Planck Institute for
Radio Astronomy

XXI   Jodrell Bank Mark IA radio telescope, diameter 250 ft

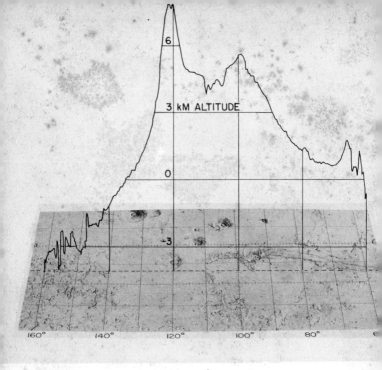

XXII

XXII Radar measurements of the planet Mars. The measured height
of the surface along the equator is shown superposed on photographs
from the Surveyor satellite (Jet Propulsion Laboratory)

An H II region may be a complicated affair, like the Orion nebula shown in Plate VII, or it may be simple and compact. A feature which all the H II regions have in common is that they all contain at least one very hot star of the so-called 'early' types O and B (see page 41). The nebula in Orion owes its complexity to the large number of these stars. The temperature of H II regions is generally rather higher than that of the sun, ranging from 5 000 K to 10 000 K. There is no problem in explaining the existence of these high temperatures, since the O and B stars which heat the gas are considerably hotter than the sun. There is, however, an interesting astrophysical problem in explaining why the temperature is not even higher, and why there is such a sharp boundary at the edge of an H II region. It turns out that there is a kind of 'safety valve' which operates at about 10 000 K and prevents the gas getting any hotter. The safety valve is oxygen, present in small proportions, which radiates strongly only when the temperature is high enough for it to be ionized. It then radiates, producing a characteristic green line in the spectrum, keeping the temperature down to about 10 000 K. This effect produces a uniform temperature out to a distance where the ultra-violet light from the star can no longer keep the hydrogen ionized, and the boundary of the cloud is reached.

There are some particularly interesting H II regions near the centre of our galaxy. The map of Figure 24 shows a non-thermal source called Sagittarius A right at the centre, with a string of H II regions on either side. These regions contain some rather more complex molecules than hydrogen and oxygen (Chapter 9).

The Hydrogen Line

Synchrotron radiation and thermal radiation from H II regions account for almost all of the radio energy emitted by the Milky Way, but they account for only a small fraction of the available radio-astronomical information on the Milky Way. The most important part is played by the 21-cm-wavelength spectral line of neutral hydrogen, which has provided some remarkably detailed and complete information on the structure and rotation of our galaxy.

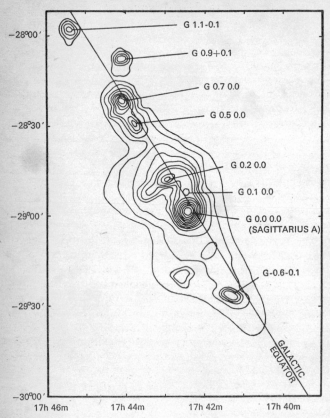

24. The radio sources at the centre of the galaxy. The source Sagittarius A is believed to be the active nucleus of our galaxy. The weaker sources are ionized hydrogen regions (after D. Downes and A. Maxwell, Haystack radio telescope of MIT at 3·7 cm wavelength)

The discovery of the hydrogen line came through the first clear application of an optical astronomer to the new radio field. During the years of the German occupation of Holland, the Leiden Observatory was directed by Professor Oort. But there was little work to direct in a practical field, and Professor Oort was mainly concerned to keep discussions and theoretical work alive amongst the few students who had escaped detention. Among these was H. van de Hulst. The idea of using radio waves to extend the study of the universe was discussed several times, but Professor Oort maintained that radio was a very weak and blunt astronomical tool compared with the optical spectroscope. How could distances, velocities, and temperatures be measured unless a spectral line could be observed? So he set Van de Hulst the task of finding, by theory alone, if any of the common constituents of interstellar space might be found to radiate a spectral line in the radio range of wavelengths.

Spectral lines are emitted by a discrete quantum process, in atoms or molecules. The energy of the emitter can only assume certain levels, and a change from one level to the other, involving a change of energy E, is accompanied by the emission or absorption of energy at a frequency ν, where $E = h\nu$, and h is Planck's constant. For visible spectral lines the change E is of the order of five electron volts, or about 10^{-11} ergs, so that for radio waves, where the frequencies are about 10^6 less than light waves, the energy change must be only 10^{-17} ergs. The problem then is to find an atom or molecule which not only has two energy levels only 10^{-17} ergs apart, but which could be expected to exist at these levels when the element was found in the cold spaces between the stars. Van de Hulst was able to show that the unionized hydrogen atom, called by spectroscopists the H I atom, had a pair of energy levels separated by this amount. The 'ground state', or lowest energy level, in which a hydrogen atom would normally be found in interstellar space, was in fact split into two levels by a magnetic coupling between the proton nucleus and the single electron. A close analogy is obtained by considering both proton and electron as weak magnets, constrained to lie parallel to one another and magnetized either in the same direction (parallel) or in opposite directions (anti-parallel). The parallel situation has a very small excess of energy over the anti-parallel situation; one can picture the magnets flipping over to the anti-parallel case, using up stored energy in doing so.

Having found the energy levels, the next stage is to determine the probability of a transition occurring from one level to another. It turns out to be an unlikely occurrence – left to itself, an excited, or parallel atom would flip over and emit its quantum of radio energy only after some millions of years. In the galactic plane the emission is stimulated by collisions with other atoms; even this only decreases the life-time of an excited atom to fifty years. But there are very many hydrogen atoms in the disc of the galaxy, and Van de Hulst was able to show that even this small transition probability gave a fair chance that the hydrogen line would be detected.

Not long after the war, as part of an independent atomic-physics research programme, physicists in the USA showed that it is possible to stimulate the transition in a beam of hydrogen atoms, and thereby to measure the radio frequency of the spectral line. This was measured, with great precision, as 1 420·403 MHz, giving a wavelength of just over 21 cm. (The wavelength of sodium light, for comparison, is about 6×10^{-5} cm.)

At the end of the war Leiden Observatory was able to publish this new idea, and to start experiments to detect the 21-cm spectral line in galactic radiation. These experiments met great technical difficulties, which culminated in a frustrating conflagration in which the whole receiver was destroyed. The credit for the first detection of the line went instead to Harold Ewen, then a research student at Harvard. Publication only came when his results had been disclosed to Leiden, and to the Australian observers, Christiansen and Hindman, who with commendable speed managed to improvise equipment and confirm the result within two months of the first discovery.

The Doppler Effect in the Hydrogen Line

The strength of the hydrogen line varies markedly over the sky. Like the synchrotron radiation, it shows a sharp concentration towards the plane of the Milky Way, indicating that neutral hydrogen is concentrated in a thin flat disc whose thickness is only about one hundredth of its diameter. Far more detail than this is available, however, since the Doppler effect enables us to measure the velocity of the hydrogen, and from this we can infer its distance.

The Doppler effect is familiar in everyday life as a change in pitch of moving sources of sound, such as sirens on passing automobiles. In optical astronomy it has long been used to measure the radial velocities of stars through the wavelength shift in spectral lines, and in the 'red shift' of the light from distant receding galaxies (Chapter 10). For the hydrogen line it is possible to measure a frequency shift of only a few kilohertz, corresponding to a velocity of about 1 km sec^{-1}; velocities of over 200 km sec^{-1} are often found in the Milky Way, so that the velocities of hydrogen clouds are easily and accurately measurable. The only limit is provided by the random motion of the gas within the cloud itself, which broadens the radiation from an extremely narrow line-width to the usual width of several kHz (see Figure 25).

In any direction in the galactic plane we look out through a disc containing a series of spiral arms radiating out from a central concentration of stars. It has been known for some time that this disc is rotating, and, further, that it does not rotate as a solid disc, but with a motion like that of our solar system, where each planet pursues a course round the sun with a velocity depending on its distance from the sun. The result is that a line of sight in the galactic plane must contain parts of the plane which appear to move towards or away from us, and this movement will cause the spectral line of the hydrogen, in this direction, to show a corresponding spread in frequency. Given the law of galactic rotation, any frequency component of this spread line can be directly related to a definite point in the galactic plane.

Take for example the spectral line shown in (d) of Figure 25. This is an actual example of the H I spectrum from a direction in the galactic plane. All the deflections come from H I radio waves, but there are seen to be three groups, one at the frequency of the undisturbed line radiation, and the other two shifted in

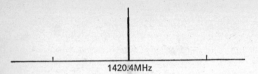

(a) Ideal hydrogen-line spectrum

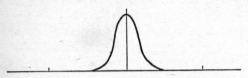

(b) The spectrum is broadened by the random motions within a hydrogen cloud

(c) The cloud as a whole is receding with a velocity of 100 km/sec.

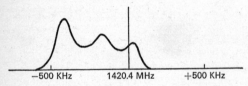

(d) In general the profile is made of components from separate clouds with different velocities.

25. The profile of the spectral line emitted by interstellar hydrogen at a frequency of 1 420 MHz (wavelength 21 cm)

frequency. The line of sight along which the radio telescope was directed when this spectrum was taken therefore must cut through three separate concentrations of hydrogen, one stationary and the other two receding from us. As the sun is itself a spiral arm, it seems likely that the stationary hydrogen is in this same local

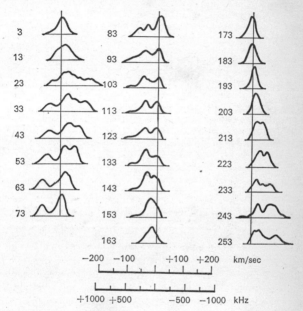

26. Profiles of the 21 cm line radiation from hydrogen at various longitudes in the galactic plane (from the work of Leiden Observatory)

arm, the moving clouds being more distant. We shall see that they are in two arms lying further out from the centre of the galaxy than the local arm.

Figure 26 shows similar profiles for the spectral line from directions all round the galactic plane. All the line profiles have a distinct maximum near the natural frequency of the line, indicating that hydrogen at rest, or nearly so, is to be seen in all directions round us. This is the hydrogen of the spiral arm in which we are situated. There are also extensions of the profiles in the directions both of receding and approaching velocities, indicating more distant hydrogen, and these parts of the profiles also contain distinct peaks.

Taking as a key this identification of Doppler shift with distance the profiles can now be interpreted into a map of the hydrogen in the galactic plane. For first approximation it need only be a two-dimensional map, since the hydrogen is mostly found within about 300 light-years of the plane, while the extent of the plane

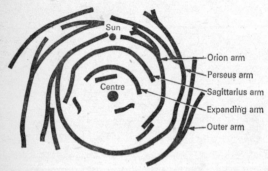

27. Spiral arms in our galaxy. The map is incomplete beyond the centre of the galaxy, but a rough spiral pattern can be traced. The arms trail due to the clockwise rotation of the whole pattern (after F. Kerr)

is more than 40 000 light-years across. Such a map is found in Figure 27 where the major concentrations of hydrogen are shaded in. Several series of observations, from both hemispheres of the earth, are combined for this map, since no single observatory can see and map the whole sky.

The Spiral Structure of the Galaxy

The map in Figure 27 has the obvious defect of not showing us a neat and tidy series of spiral arms, radiating uniformly from the centre of the galaxy, but a glance at other spiral nebulae, Andromeda for example (Plate I) or M51 (Plate II), shows that this is hardly to be expected. The appearance is more of a whirlpool, in which each part of the surface moves in a circle, twisting the ripples into a spiral form. In our galaxy the irregularities have been drawn out almost into a circle. Nevertheless a spiral form, with the arms trailing, is still discernible, showing that the irregularities have been dragged about four times round by the differential rotation. The spiral is tightened a little more each time the whole disc makes one rotation about the centre.

Does the galaxy go on winding itself up tighter and tighter? The answer must be no, as in its lifetime so far it should have been wound up about ten times as much as it now is. There must be some process of re-forming at work, in competition with the rotation, a process which would be a departure from the regular circular motion. Of this process we know very little, except that it seems to be a consequence of a dynamical wave-like motion of the stars and interstellar gas, following paths differing slightly from a true circular motion.

The Magnetic Field of the Galaxy

One further constituent of interstellar space has a particular importance in radio astronomy. This is the magnetic field, whose existence has already been implied in the attribution of the major part of galactic radiation to the synchrotron process. The strength of the field cannot be greater than a few millionths of the magnetic field at the surface of the earth, but its effects are nevertheless easily noticeable in motions of ionized gas clouds in the Milky Way. The field seems to be fairly smooth and well-organized; this is revealed by the polarization of the synchrotron radiation (see Chapter 2), which is quite high in some parts of the sky.

The major contribution of radio astronomy to the study of this field has been an actual measurement of its strength. This has been achieved through the Faraday effect, which is the rotation of the plane of polarization of a wave passing through a transparent medium in which there is a magnetic field along the direction of propagation. Faraday first demonstrated the effect with light passing through a block of glass in the field of a powerful electromagnet, but it has been studied most as an effect on radio waves within the terrestrial ionosphere. Here, as in interstellar space, the effect depends on the ionized part of the gas which is traversed by the radio waves. Figure 28 shows how the plane of polarization of a radio wave rotates as it traverses a spiral arm of the Milky Way. The actual rotation angle depends on the product of total electron content and the magnetic field along the propagation path; it also varies as the square of the radio frequency, so that it can be measured by observing the apparent plane of polarization of the same source at several different radio frequencies.

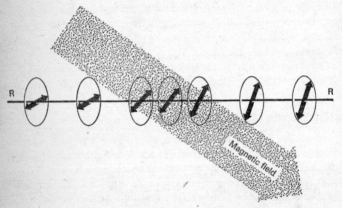

28. Faraday rotation in a spiral arm. The plane of polarization along a wave RR is rotated by the magnetic field in the ionized gas of the spiral arm

Faraday rotation has been observed in many radio sources, particularly in extragalactic sources, but the measurement gave only the product of electron content and strength of field. Sorting these two out only became possible with the discovery of the pulsars, and the discovery that their radiation is very highly polarized. This immediately led to a measurement of the Faraday rotation along the line to several of the pulsars. The particular importance of this measurement was that the total electron population along the line was already known from measurements of pulse delay (Chapter 7). The separated value of the magnetic field was then found for pulsars over a range of directions in the plane of the Milky Way. The direction of the field, along the local spiral arm, and its strength, about 3 microgauss, were thus obtained for the first time.

Irregularities in the Interstellar Gas

The ionized interstellar gas can be observed through several radio methods. If it is sufficiently dense, as in an H II region, it may radiate thermal radiation or, at low frequencies, it may absorb radiation from more distant sources. As found in pulsar studies, it can delay the arrival of radio pulses, and with the magnetic field it can rotate the plane of polarization of radio waves from quasars and pulsars. All these occur from smoothly distributed gas. There is a further effect which depends not on the smooth distribution but on irregularities in the gas density. This is a scintillation effect, like the familiar twinkling of visible stars. An example is shown in Figure 29, which shows the fluctuations of signals from a pulsar over an observing period of an hour.

Scintillation is used to study the irregularities in the ionized interstellar gas, which are believed to be in the form of waves travelling through the gas. It is a similar phenomenon to the scintillation of radio sources seen through the outer corona of the sun, already described as interplanetary scintillation (Chapter 4). Interstellar scintillation has also been used in measurements of the velocity of pulsars through the interstellar gas.

We turn now to some of the denser irregularities in the interstellar gas, where the thinly distributed atoms seem to be concentrating into discrete clouds. At a sufficiently high concentration the atoms start to cohere, forming molecules and dust particles, the first stages of stellar formation. These concentrations may contain enough mass for hundreds of stars, and they provide a rich variety of radio radiations which reveal some of the conditions in the birthplace of stars, and possibly also the cradle of life itself.

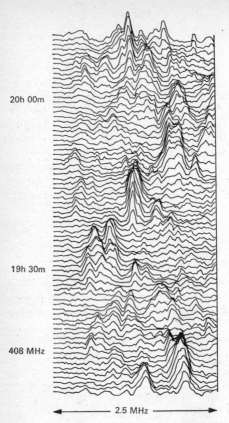

20h 00m

19h 30m

408 MHz

◄——————— 2.5 MHz ———————►

29. Scintillation of radio waves from the pulsar PS R 0329 + 54. Each profile represents a spectrum, measured at time intervals of 50 seconds. A band 2·5 MHz wide at 408 MHz is depicted. The scintillation is caused by irregular ionized gas clouds in interstellar space (after B. J. Rickett, Jodrell Bank)

CHAPTER 9

Molecules in Space

The Origins of Life

IN recent years tremendous advances have been made in the physics and chemistry of living matter. We no longer think of a vague jelly-like material, protoplasm, out of which living things are constructed; instead we know, at least in outline, the chemical structure and physical configuration of many of the important animal-protein substances. We know also how genetic information is passed on to new individuals of the same species, and how variations in that information can lead to variations in individual characteristics and to the development of species. There is limitless detail remaining to be unravelled in biochemical and biophysical studies, but there is now a firm understanding of the basic structure and the evolutionary history of living things. These are seen in terms of large molecules – with tens of thousands of atoms in each, composed mainly of carbon, hydrogen, and oxygen – whose structure shows almost infinite possibilities of small and subtle variations.

There has often been speculation about the contrast between the very high degree of organization in living material and the general tendency towards disorder which is a natural physical law of the whole universe. The organization of large and complex molecules and their construction from their elementary inorganic constituents is a process which the human mind finds it difficult to comprehend in all its detail, and philosophers concerned with the origin of life have often been constrained to postulate the intervention of a life force, or spiritual agency, which can act in contravention of normal physical laws.

The misunderstanding leading to this postulate has been exposed by J. Monod, the Nobel laureate who is primarily a biochemist but incidentally also a philosopher. The point is that the amount of organization required for the construction or modification of a biological molecule is comparatively low. If physical laws are to be involved in the argument, then the organization must be measured by the energy required to construct a large organic molecule out of small inorganic molecules, such as water and carbon dioxide. This energy is in fact low in comparison with the energy required for the construction of water and carbon dioxide out of their constituent atoms of hydrogen, oxygen, and carbon. The energy required in the mutations of genetic messages which are responsible for evolution is much lower again. If we are to speculate about the way in which life came to exist in the universe, our first concern must therefore be with the large energy steps which are involved in the construction of small molecules. Given these basic building materials, there is a good chance that life can exist on planets in many different parts of the universe, depending mainly on a delicate balance between factors such as the rate of solar heating, the density of the atmosphere, and the rate of condensation of the planets from the nebulous material surrounding a star.

The circumstances in which biological evolution started on earth may therefore also be present in many other locations within the Milky Way. Some adventurous observers have already devoted their efforts to a search for intelligent life in the universe by looking for evidence of extraterrestrial intelligence in cosmic radio signals. More soberly, and more profitably, it has now been found possible to demonstrate the existence of simple molecules in interstellar space, and to study the circumstances of their formation. This is achieved through radio spectroscopy, the study of the radio spectral lines of more complex substances than hydrogen, whose 21-cm-line radiation was discussed in Chapter 8.

The Search for Deuterium

After the success of the 21-cm hydrogen line, it seemed that the first step should be a search for the heavy isotopic form of hydrogen, known as deuterium. On earth there is about one atom of deuterium to every four thousand of normal hydrogen, so that the main spectral line, known to be at 92-cm wavelength, was expected to be weak but just detectable. Searches were made in various parts of the Milky Way, especially near the centre of the galaxy, but without success. This initial failure was a discouragement to radio spectroscopy, and probably contributed to a long delay between the discovery of the hydrogen line and a prolific age of about ten years which contained a flood of discoveries.

The Hydroxyl Maser

The turning point for molecular spectroscopy came in 1963 with the discovery of strong spectral-line radiation from a very basic substance called hydroxyl, the OH radical. This part-molecule has four closely spaced spectral lines near 18-cm wavelength. These lines have been observed both in absorption and in emission from many locations close to the plane of the Milky Way, giving clear proof of the existence of one of the basic building bricks of organic chemistry.

The first detection was in the form of an absorption line. The depth of the absorption of radiation from a powerful radio source, such as Cassiopeia A, should give a measurement of the amount of OH along the line of sight to the source. As with the 21-cm hydrogen line, a precise measurement of the radio frequency of the absorption line can also give the distance. The discovery of the OH lines, in 1963, at first gave a fairly straightforward picture of OH concentrated in the Milky Way. There was, however, a small puzzle in the ratios of the strengths of the four absorption lines, which did not agree with theory, but this was hardly enough to upset the simple interpretation of the first observations.

This simple picture was shattered as soon as the radio spectrometers were set to look for OH-line emission. Very strong lines were found emitted by H II regions, such as the Orion nebula, far stronger lines and much narrower than any that had been seen from hydrogen. Figure 30 shows the complex nature of OH-line radiation from the H II region W3. There are several line components, and they are circularly polarized. These lines could not be the natural line radiation; they could only be the result of an amplification process only recently discovered in physics laboratories and known as the 'maser', or 'microwave amplification by stimulated emission of radiation'. For this process energy is fed into the OH gas, for example by absorption of ultra-violet light, and re-appears through quantum processes as an amplification of the natural line radiation. (It is closely related to the better-known laser, which relates to light rather than microwaves.) The existence of maser amplification in OH completely upset the calculation of OH concentration in the galaxy; it did

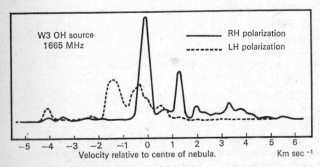

30. Hydroxyl spectral line emission from the W3 nebula. The emission is centred on 1 665 MHz; components of the line are radiated by parts of the nebula moving at various velocities relative to the nebula's centre, over a range of 10 km sec^{-1}. Different components are seen for the two hands of circular polarization (250-ft Mark I radio telescope Jodrell Bank)

however account for the odd line ratios seen in the four absorption lines.

A fresh approach to the calculation of OH concentration came through attempts to measure the diameters of the sources of OH-line emission – they were soon found to be small, concentrated objects embedded in or close to the H II regions. A further advance came in 1968 when W. J. Wilson and A. H. Barrett in the USA observed OH emission lines from the atmosphere of a star, the red supergiant NML Cygnus.

These OH sources are now thought to be gas clouds which are condensing to form stars. This condensation is a necessary stage in the evolution of a galaxy from a collapsing gas cloud, but it seems to be a continuing process in the dense gas clouds near the centre of the Milky Way. The process depends critically on gas density, since a minimum concentration is needed for further condensation to start. When it does start, the increased density prevents starlight from penetrating into the cloud, and the interior becomes cooler, and consequently the condensation proceeds further. The end product is a star, but in the early protostar period, at a particular stage in the process of condensation, the conditions are ideal for molecules to form. It is here that the celestial biosynthesis begins, that the OH radicals are produced, that the first steps are taken towards the organization of the elements – towards life.

Ammonia, Water, and Formaldehyde

With the demonstration of the existence of OH in vast quantities, the probability of the existence and detectability of other molecular species came to dominate the plans of radio astronomers. A new subject, cosmochemistry, was born, combining the efforts of spectroscopists predicting the radio frequency of spectral lines, chemists predicting the rate of molecular formation in protostars, and radio astronomers extending their apparatus to work at the shorter wavelengths, where most of the lines appear.

A radio spectral line was already known to exist for ammonia; it had been demonstrated in the laboratory, even including maser amplification. This was found from the celestial protostar sources, but not with maser amplification. Water vapour was well known to emit a line at 13-mm wavelengths; the terrestrial atmosphere both radiates and absorbs at this wavelength. Surprisingly, there is water vapour in the Milky Way, and particularly in the atmospheres of cool red stars; even more surprisingly, its line radiation can easily be detected through the water in the terrestrial atmosphere. This is only possible because the celestial water line is amplified and narrowed by maser action. (Hydroxyl and water are the only species known to have maser amplification.) Figure 31 shows the line radiation from water vapour in the Orion nebula.

Moving further towards the vital components of organic

chemistry, several carbon compounds have now been detected, starting with the simple carbon monoxide, CO. This can be detected from most of the galactic plane, at a wavelength of 2 mm. (Here, incidentally, is a new source of information about galactic dynamics; since the CO line has a natural width small in comparison with the Doppler shifts which occur due to galactic rotation, it should be very profitable to use it to improve on the hydrogen-line studies of spiral arms.) The final step across the

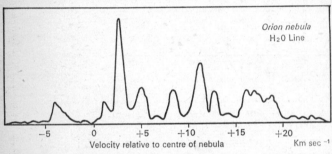

31. Spectral line emission from water vapour in the Orion nebula. The line is centred on 22 235 MHz, at 1·35-cm wavelength. Components of the line are radiated by parts of the nebula moving with a range of velocities (from the Naval Research Laboratory, Maryland Point Observatory)

threshold into organic chemistry came with the discovery of a line from formaldehyde, H_2CO.

The Complex Molecules

The commonest atomic species in space is hydrogen. Oxygen is also fairly abundant, while atoms such as nitrogen, carbon, sulphur and silicon are known to be present but less abundant. Simple compounds with hydrogen, such as OH and NH_3 are therefore the first to be expected, while those such as formaldehyde with both carbon and oxygen are less likely. The list of successful observations in the Table on page 131 is therefore remarkable in the number of compounds now observed which have two or more of the heavier and rarer elements combined.

SPECTRAL LINES DETECTED UP TO 1972

Year of discovery	Species	Symbol	Wavelength (centimetres)
1957	Atomic hydrogen	H	21
1963	Hydroxyl	OH	18
1964	Recombination lines	H, He, etc.	—
1968	Ammonia	NH_3	1·3
1968	Water	H_2O	1·4
1969	Formaldehyde	H_2CO	6·2
1970	Carbon monoxide	CO	0·26
1970	Cyanogen	CN	0·26
1970	Hydrogen cyanide	HCN	0·34
1970	X-ogen	—	0·34
1970	Cyano-acetylene	HC_3N	3·3
1970	Methyl alcohol	CH_3OH	36
1970	Formic acid	CHOOH	18
1971	Carbon monosulphide	CS	0·20
1971	Formamide	NH_2CHO	6·5
1971	Silicon oxide	SiO	0·23
1971	Carbonyl sulphide	OCS	0·27
1971	Acetonitrile	CH_3CN	0·27
1971	Isocyanic acid	HNCO	1·4
1971	Hydrogen isocyanide	HNC	0·33
1971	Methyl-acetylene	CH_3C_2H	0·35
1971	Acetaldehyde	CH_3CHO	28
1971	Thioformaldehyde	H_2CS	9·5
1972	Formaldimine	CH_2NH	5·7

Note for example the presence of carbon monosulphide, CS, and of acetaldehyde CH_3CHO, which contains seven atoms. The wavelengths shown in the Table are those of the principal spectral lines; most of these molecules radiate several detectable spectral lines at radio wavelengths.

The third entry in the Table refers to a series of spectral lines which occur through the whole radio spectrum. These are the 'recombination' lines of ionized hydrogen, and to a lesser extent

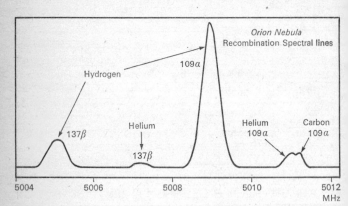

32. Spectral lines from elements in the Orion nebula. The labelled features are in the 'recombination' lines from atomic transitions in hydrogen, helium, and carbon (from Churchwell and Mezger, NRAO 140-ft radio telescope)

of heavier elements (see Figure 32). Recombination lines are very similar to the optical line radiations of hydrogen, but they are the result of quantum changes in electron orbits far from the nucleus instead of close to it. (Quantum numbers of over 200 may be involved.)

A rather mysterious entry appears after hydrogen cyanide at a wavelength of 3·4 mm. This is a line which so far is unidentified, and called X-ogen until it can be assigned to yet another molecular species.

The dates of these discoveries show that 1970 and 1971 were golden years for cosmochemistry. This was the direct result of the construction of a new radio telescope at Kitt Peak in Arizona. This 36-ft reflector works at millimetre wavelengths, where most of the discoveries were made. Other lines are still to be found at longer wavelengths, however, and particularly those from the larger and more complex molecules, as for example acetaldehyde at 28-cm wavelength and thioformaldehyde at 9·5 cm.

The search for deuterium may now be recalled as the first search for an isotopic species. Heavy hydrogen itself still has not been found, but along with other isotopes it has been detected when combined with other atoms. Deuterium is found in deuterium cyanide, DCN. Similarly, the isotopes of oxygen and nitrogen are both detected in modified spectral lines from isocyanic acid HNCO. The relative abundances, as seen in clouds near the centre of the galaxy, are the same as on earth.

The formidable list of molecules in the Table, together with their various isotopic species, might suggest that it only requires a chemist to describe a species and compute its radio spectral lines for its discovery to be virtually certain. There are, however, very many exceptions, the most notable of which is the radical CH, which ought to be as easily formed as OH but which has not so far been detected. Here some attention must be given to the processes both of formation and of destruction; CH is so reactive that it does not exist for long without combining with other atoms or molecules to form more complex molecules, and the concentration of CH may never build up sufficiently for its detection.

The Formation of Molecules

The various molecular species which have so far been detected through their radio spectral lines are all concentrated towards the plane of the Milky Way, and concentrated also towards the denser part of the Milky Way near the centre of the galaxy. Some, such as carbon monoxide, are found widely distributed along the plane, but most, including the more complex, are concentrated into small sources, which are the dense protostar clouds. Some of these clouds contain an identifiable star in the early stages of formation, when most of the light is emitted at the infra-red end of the visible spectrum. In these protostar clouds the particle density is possibly 10^{10} times larger than that of free space, and there is a reasonable chance that atoms can collide, forming radicals and molecules.

At the same time there is no intense starlight, particularly ultra-violet light, to disrupt molecules by direct photon action or by heating the whole gas. Even the energy supply for the maser-amplified lines must come only from the atomic collisions, and not from starlight.

Here, it seems, must be the cradle of life in the Milky Way. The presence of many more complex molecules will probably be detected in the dense protostars; the most complex to date are thioformaldehyde, containing sulphur instead of oxygen, and formaldimine, which notably contains both nitrogen and carbon. But there is a remaining problem – how do the molecules leave the protostar? If they stay inside, their fate is certain. As the star condenses, it must heat up and destroy all molecular species by its light emission. On a simple picture, the collapse and the destruction is inexorable. But a simple picture of star formation is not enough to account even for the formation of our own planetary system round the sun, so we can only reserve judgement on the fate of these newly born molecules. One thing is, however, now quite certain: organic molecules, from which living molecules can easily be constructed, can be formed by natural processes, and exist in the same form in interstellar space as they do on earth. Interstellar formic acid is the same compound as can be distilled from ants: interstellar formaldehyde would serve well for preserving biological specimens, and interstellar water would be perfectly drinkable, if enough could be collected together to make a mouthful.

CHAPTER 10

Galaxies

THE first large radio telescope built at Jodrell Bank was a parabolic reflector 218 ft in diameter. It was a spider's-web structure, made of scaffold poles and wire, fixed to the ground and directed vertically upwards. It was not originally intended for use in radio astronomy, but for the quite different purpose of detecting cosmic-ray showers by radar. In this purpose it failed: the detection of cosmic-ray showers by radio was in fact first achieved at Jodrell Bank many years later by quite different means. The radar experiment itself was abandoned, but instead the reflector was equipped with a sensitive receiver, and, as a radio telescope, was used to make a radio map of the part of the sky crossing the zenith at Jodrell Bank. The next stage was to cover more sky by swinging the telescope beam north and south of the zenith. This brought into view the Andromeda nebula, the best-known of all the extragalactic nebulae, and made possible the first recordings of its radio emission.

As we shall see later in this chapter, two other extragalactic nebulae were at this time already known to be strong radio sources. These two are, however, now thought of as 'radio galaxies', a class of galaxies in which there is some form of violent commotion leading to enhanced radio emission. The Andromeda nebula was the first 'normal' galaxy to be detected. It is a spiral galaxy, and the largest member of the local group of galaxies in which our own galaxy, the Milky Way, is situated.

The Andromeda nebula (Plate I) is seen at an inclined angle, so that the circular disc appears elliptical. The overall pattern of this disc can be seen with much greater detail than we can hope to see in our own galaxy. In the spiral arms powerful telescopes can distinguish the bright O and B stars, many with ionized gas surrounding them. The fainter stars extending through the whole nebula are mostly unresolvable, but show as a continuous glow like our Milky Way, enabling us to delineate the shape of the nebula by photometric measurements of long-exposure photographs.

What can radio astronomy add to the beautiful detailed pictures that optical telescopes have already obtained of the Andromeda nebula and the other nearby galaxies? There are two ways in which radio can help; these refer to the synchrotron radio continuum and to the 21-cm line radiation respectively.

Synchrotron Radiation from Normal Galaxies

Radio emission can be detected over a wide range of wavelengths from normal galaxies, as it can from the Milky Way. Its strength depends on the population of energetic electrons and on the magnetic field of the galaxy, neither of which can be detected optically. This radiation can be detected from all types of normal galaxies, including the irregular galaxies such as the Magellanic Clouds, and the elliptical galaxies which have no spiral structure. So we first deduce that electrons and magnetic fields are to be found in galaxies other than our own.

More detail of the radio emission from the local galaxies has now been obtained, using the new high-resolution 'aperture-synthesis' radio telescopes such as that at Westerbork in the Netherlands. Plate II shows the spiral nebula M51, with contours of radio brightness superposed on the photograph. Now we see that the radio emission delineates the spiral-arm structure, but with a tendency to be strongest just outside the optically brightest parts. This is supposed to be a region where the magnetic field is compressed, giving increased synchrotron emission. This is an important observation in the theory of formation of spiral arms, which are now thought to be the result of a wave-like motion travelling in the disc of a galaxy. The compression of the magnetic field is a natural consequence of the wave motion.

The source of the cosmic-ray electrons in these galaxies, and in our own, is still unknown. There must be a considerable source of energy, associated either with the nuclear energy in stars or with the gravitational energy of a shrinking galaxy. In our own galaxy the most obvious source is nuclear energy, which makes the stars shine. The requirement of the cosmic rays is for less than one hundredth of this supply; nevertheless, there is good reason to suspect that they have quite a different origin. This will appear later in this chapter under 'Radio Galaxies', but a hint is found in several of the new radio maps of galaxies which are usually classified as normal. On these maps there is often a sharp bright peak of radio emission at the centre of the galaxy. This may be the first sign of abnormality, but it may prove to be so common that a completely 'normal' galaxy will become the exception.

Rotation Curves and Masses

The 21-cm hydrogen-line emission from galaxies is particularly informative because of the Doppler frequency shifts which provide a measure of the velocity of various parts of the galaxies. Even without the good angular resolution of an aperture-synthesis telescope, a simple analysis of the hydrogen-line emission from a galaxy can give a measurement of the mass of neutral hydrogen, the total mass of stars and gas, and the rotation-speed of the galaxy.

The Andromeda nebula, M31, is a spiral galaxy which has been examined in this way. Figure 33 shows how the velocities derived from the hydrogen line are used to find the 'rotation curve' for this galaxy. Here the rotation-speed is shown increasing with distance from the centre, out to a distance of about 4 light-years. Beyond this distance the rotation slows down, and the speed shows little increase. The shape of these rotation curves can be related to the mass distribution in the galaxy. Deviations from smooth curves, seen particularly in the inner parts of galaxies, apparently arise from radial motions of hydrogen gas, as though some of the gas is expanding from the centre.

There are other radio spectral lines which will provide new information on the extragalactic nebulae. The hydroxyl radical has already been detected in several galaxies, and it should be possible to map its distribution through the spiral-arm structure of such galaxies as M101 and the Andromeda nebula. If carbon monoxide can also be detected, then much greater precision will be obtainable in the rotation curves, particularly near the centres where the interesting non-radial motions have been found.

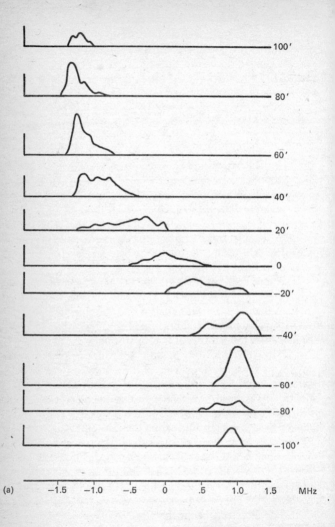

(a)

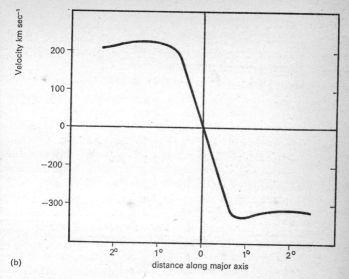

(b)

33. Rotation of the galaxy M31, the Andromeda nebula. (a) The succession of hydrogen-line profiles along the major axis of the galaxy shows velocities towards the earth on the left, and away on the right. The line profiles were made at 20'-arc intervals (250-ft Mark I radio telescope, Jodrell Bank); (b) Rotation curve for M31. The velocities along the major axis, derived from the spectra of Figure 33(a)

The Expanding Universe

The 21-cm hydrogen line gives the velocity of the hydrogen within a galaxy through the Doppler shift of the line frequency. This spreads the line over a range of frequencies, indicating the range of velocities within the galaxy. The centre of the velocity spread refers to the mean velocity of the whole galaxy, which is generally a velocity away from earth.

The first indication that galaxies are, as a whole, receding from us came from the observation by V. M. Slipher that most had optical spectra showing lines shifted towards the red. This direction of the Doppler shift is associated with motion away from the observer. In 1929 Hubble showed that the amount of

139

the red shift in the spectra of galaxies was proportional to the distance. It so happened that cosmological theory had already predicted that there should be a velocity of recession proportional to distance (see Chapter 12), so this very remarkable result was accepted fairly easily, and became the foundation of modern cosmology. There were, however, some doubts; might there be another physical explanation for the red shift, so that the inference of recession had no foundation? Perhaps light travelling long distances slowly lost energy, so that light quanta became less energetic, or more red, as they travelled through space?

The test of these theories was to look for the red shift of radio waves, since any effect of quantum energy loss should be very different at radio as compared with light wavelengths. The experimental result is shown in Figure 34, where the graph compares all the known 21-cm red shifts with the optical red shifts. The red shifts agreed within 1 per cent, giving a decisive verdict in favour of the Doppler explanation. Very little more has since been heard of the alternative explanations, and the Doppler velocities are now almost, but not quite, undisputed.

The Problems of Stefan's Quintet

Not all galaxies have precisely the velocities of recession which would be expected from their apparent distances. On a small scale, there are random movements within any cluster of galaxies. For example the Andromeda nebula is so near that it should only have a small velocity of recession; in fact it is moving towards us with a velocity typical of the random movements within our local cluster. On a large scale there are also some individual problem cases. We shall see later that some galaxies are blowing apart, so that some pieces are moving at thousands of kilometres per second away from others, but the present concern is the discrepancies that occasionally are found for whole galaxies. Of these the best known is a member of the group known as Stefan's Quintet.

The Quintet is such a closely spaced group that all five members are usually reckoned to have a common origin. But when their optical red shifts were measured, four were found to be identical and in agreement with the Hubble Law. The fifth was totally

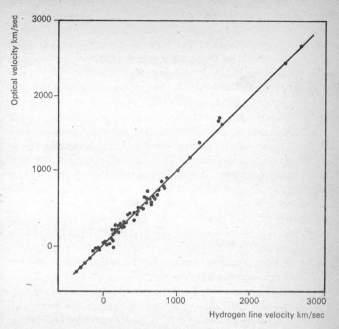

34. Radio red-shift velocities (horizontal axis) compared with optical red-shift velocities (vertical axis). Each point represents a galaxy. Within the accuracy of the observations all the points lie on a straight line with the slope of unity (from W. H. McCutcheon and R. D. Davies)

different, indicating a speed of over 1 000 km sec^{-1} relative to the other four. This difference seemed so unlikely that the more adventurous of the theorists started to reopen the question of the origin of the red shift. Could it be due to some local conditions peculiar to this one galaxy, for example a very strong local gravitational field? The arguments were important, since they came at a time when large red shifts were being found for quasars (Chapter 11), and a demonstration of any new origin for the red shift would have undermined the new results just emerging in the new field of radio cosmology.

The explanation of the odd Quintet member seems to be that

this is a galaxy which does not really belong to the others, but which is much closer and only seen as a member of the group by a chance superposition. The red shift of this odd-man-out is in fact similar to that of some other galaxies quite nearby; nevertheless a suspicion of trouble remained.

Again the 21-cm-line velocity measurements have acted as a further test. Only the Doppler explanation should produce the same velocities both for optical and radio measurements; this was found experimentally to be the case, and the expulsion of the fifth member of the Quintet was confirmed. He is, perhaps, to be regarded as a visiting soloist to a closely integrated Quartet; certainly he no longer spreads disharmony into cosmological physics.

RADIO GALAXIES

Surveys of Radio Sources

The first survey of celestial radio sources was that of John Bolton, in Sydney. He was building on previous work by Reber and by Hey, but his radio telescope was the first to be designed especially for discrete sources. The word 'radio telescope' in these days tends to suggest to us a large parabolic reflector, possibly because this type of instrument bears some resemblance to an optical reflecting telescope. Bolton's instrument looked nothing like a parabolic reflector. It consisted of an array of aerials, similar to those now used for television reception, directed towards the horizon from the high cliffs at the entrance to Sydney Harbour. Radio waves from a source just above the horizon reached this array both directly and via reflection in the sea, giving the effect of an interferometer (Chapter 16). Interferometers can be made sensitive only to radio waves which come from sources with small angular diameter, so that by directing the array to different parts of the horizon Bolton was able to watch for discrete radio sources in any part of the sky rising over the horizon at Sydney. The Crab nebula (Chapter 6) was one of these, and so was Cygnus A, which was not at that time identified. Two others were, however, tentatively identified with extragalactic nebulae in the constellations of Centaurus and Virgo. These were the first radio galaxies to be discovered out of the many hundreds which are now known.

Major radio surveys were subsequently carried out both in Cambridge and in Australia, producing material for a fierce controversy over the nature of the sources and their meaning for cosmology. This controversy is concerned with the most distant radio sources, most of which are quasars, but it depended originally on the distance scales established for radio galaxies through the identifications described in this chapter.

Cygnus A

The first discrete radio source to be seen, the first extragalactic source to be identified without any doubt, and one of the most powerful yet discovered, Cygnus A is still thought of as a proto-type of the radio galaxies. On the photographs obtained by Baade and Minkowski it looks like two small galaxies in contact; indeed, as described in Chapter 3, the first interpretation was that these were two separate galaxies in collision. This was not at all a far-fetched idea; at least near-misses do occur, as we can see in several pairs and groups of galaxies which are grotesquely distorted by one another's gravitational fields. The two Magellanic Clouds, for example, show some of this gravitational distortion, while the two Clouds together distort the outer parts of the Milky Way. The idea of a collision was, however, soon dropped when a search for the difference in Doppler shift of the optical lines between the two halves of the galaxy showed no appreciable systematic velocity difference. Some other explanation was needed for the energetic state of Cygnus A.

Most of the radio emission, and often some of the light, from radio galaxies is synchrotron emission. The strength of this emission can be used as a measurement of the energy stored in a galaxy like Cygnus A. We know that the rate at which N electrons radiate power, over a wide range of frequencies, is proportional to NE^2H^2, where E is the energy of the electrons and H the magnetic field. The total energy in N electrons is NE, and the energy stored in the magnetic field is proportional to H^2. The observed synchrotron radiation might be generated by many electrons in a weak field, or by fewer electrons in a stronger field, and the total energy stored in both together is not therefore directly derivable from the strength of the radiation. But it is easily shown that the least possible value for the total energy is obtained when the energy is shared roughly equally between them. This is the condition of 'equipartition'. Even though the actual total energy may be considerably greater, the minimum figure is already astonishingly large, and theorists find it so hard to account for that they ignore the possibility of the worse dilemmas presented when equipartition is not obeyed.

Some orders of magnitude may illustrate the problem. For several radio galaxies the minimum total energy exceeds 10^{59} ergs. By comparison, the energy available in transforming a whole solar mass of hydrogen into helium is just over 10^{52} ergs, and even the total conversion of a solar mass into energy would only yield 2×10^{54} ergs. The source of energy which makes the stars shine is totally inadequate to account for the brightness of the radio galaxies, unless a considerable proportion of the mass of the galaxy has been transformed into energy.

On the photograph in Plate XI the visible nebula is about 10 seconds of arc across. If the identification had been made sooner than it was, the smallness of this diameter, which is about thirty times smaller than the smallest angular distance resolvable by the human eye, would have deterred radio astronomers from any attempt to measure the diameter of the radio source. Obviously the optical and radio diameters need not be the same, since the light comes from heavy ions in the interstellar gas and in stars, while the radio comes from electrons moving in a magnetic field which might be more extensive than the visible nebula. However, no one could have predicted the result of the first measurements which give a radio size close to 3 minutes of arc, about twenty times the size of the visible object. The true situation is now revealed in Figure 35, which is a detailed map of this astonishing radio source.

The sketch of the optical nebula in this figure may be compared with Plate XI, when it will be seen that the radio source is not on the Plate at all, but is in two extended regions on either side. The outer edges are well defined, while the inner edges shade off to-

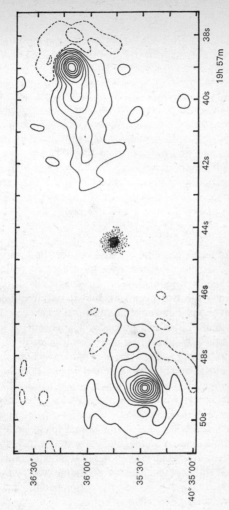

35. A radio map of the Cygnus A radio galaxy, showing the two regions of radio emission on either side of the visible nebula (Cambridge 1-mile telescope at 6-cm wavelength)

wards a region of almost no emission, in the centre of which is the visible nebula.

There was a period of almost a year during which this extraordinary difference between the optical and radio appearances of Cygnus A led many astronomers to doubt the reality of the identification. Cygnus A happens to be near the plane of the galaxy, and it was possible that by accident a supernova radio source, like Cassiopeia A, was in the line of sight. This would dispose at once of the difficulties of explaining the enormous energy requirements, and the enormous size of the radio galaxy which appeared to be about ten times the dimensions of our own galaxy. The question was only finally settled by the identification of another source, 3C 295.

Into the Depths of the Universe – 3C 295

The galactic, and extragalactic, possibilities for the location of Cygnus A were at distances differing by a factor of about a million. The red shift, only observable optically, provides a very clear test between the two distances, and it only required the identification of another powerful radio source with a red-shifted galaxy for the final doubts to be allayed. The Cygnus A galaxy had a red shift of 0·06, indicating a recession velocity of 7 per cent of the velocity of light and a distance of almost 1 000 million light-years: could another such be found in the position of a radio source?

In 1960 a very accurate position found for one of the sources in the Cambridge catalogue, 3C 295, was used by Minkowski for another optical search with the 200-in Palomar telescope. Exactly in the right spot he found another galaxy whose spectrum showed the unusual emission lines already found in Cygnus A. This time, however, the red shift was 0·45, showing that the galaxy was seven times further away again; the radio observation had tracked down the most distant galaxy ever observed at that time.

Furthermore, the shape of the source was soon found to be very similar to that of Cygnus A, only on about one tenth of the angular scale. Everything fitted well with the interpretation: there was another Cygnus A, a little more powerful, but the same general size and shape. Everything now fell into place, and for a while the term 'radio galaxy' meant only one type of object, with some variations and aberrations. The shock of the quasars was yet to come, but even amongst the ranks of the known radio galaxies there were some strangers, as will now appear.

Perseus A (NGC 1275)

Many of the visible galaxies occur in clusters, of which our local group of galaxies is a modest example. It was at one time expected that clusters might contain sufficient material filling the space between the galaxies that radio emission might be detected from the cluster as whole, as contrasted with the weak radiation from the individual members. This idea seemed to be confirmed when a strong radio source was found in the direction of the Perseus cluster, which is a prominent group about 2° across and at a distance of about 100 million light-years. There was some excitement about this discovery, as it appeared that the cluster radiated about four times as much as would be expected from a group of normal radio galaxies. However, Baade pointed out a peculiar galaxy in the centre of the cluster, and suggested that this galaxy alone was the radio source. The galaxy pointed out by Baade is NGC 1275, which proves indeed to be the strongest radio source. There are, however, other sources, including two strangely distorted radio galaxies known as IC 310 and NGC 1265, both within $\frac{1}{2}$° of NGC 1275. These two may be related to the main radio galaxy; it has been suggested that the 'tails' seen in Figure 36 are the result of a galactic wind blowing from NGC 1275.

NGC 1275 was once thought of as a prime example of a collision, since it seems to have two distinct parts which are moving at a speed of over 1 000 km sec^{-1} relative to one another. This violent motion is in fact derived from inside the nebula. Most of the radio emission is from the same general region as the visible galaxy, which is about the same size as our own galaxy, but there is also a very concentrated nucleus containing a radio source only a few hundred light-years across, which shows up in high-resolution interferometer observations.

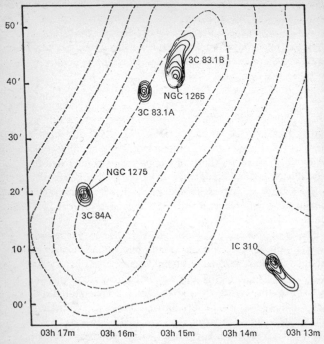

36. Radio sources in the Perseus cluster. The strongest source is the galaxy NGC 1275. The two galaxies NGC 1265 and IC 310 have trails pointing away from NGC 1275, possibly blown by an intergalactic wind. The radio source 3C 83·1A has no associated visible galaxy and is probably not part of the Perseus cluster (Cambridge 1-mile telescope)

Centaurus A (NGC 5128)

This nebula is classified 'peculiar', and it well deserves the epithet. It appears to be a nearly spherical galaxy, without spiral-arm structure, but with a dark band stretching right across it. One explanation for this dark lane is that it is another galaxy, a much flatter one, seen edge on in front of the spherical galaxy. This is a most unlikely coincidence. The strong radio emission points again to a very powerful disturbance within the galaxy. The radio

148

source is a 'double double', the outer double stretching over 4 degrees of the sky as compared with 5 minutes of arc for the inner double. This angular extent allows the possibility of detailed mapping not only of the radio brightness but also of the polarization of the radio waves, which shows up well over most of this radio source.

Virgo A (M87, NGC 4486)

This galaxy is in a class by itself. It is an elliptical galaxy of type E0, that is to say it is very nearly spherically symmetrical. In Plate III a short exposure has shown the unique feature of a bright blue streak or jet protruding from the centre. The light from this jet is polarized, just as is the light from the Crab nebula. This was the first extragalactic object found to radiate polarized light.

The radio emission from Virgo A originates in three distinct sources. There is radio emission from the jet, corresponding very closely with the polarized blue light, but with an extension across to the opposite side of the galaxy, forming a kind of 'counter-jet'. There is also an extended radio 'halo', larger than the optical galaxy. Finally there is a bright spot right in the centre of the galaxy, which appears to have a double structure very similar to that of the larger and more powerful radio galaxies, but on a very much smaller scale.

The Exploding Galaxy M82

The radio source 3C 231 was at first thought to be the large spiral M81, but accurate positions narrowed down instead on a smaller neighbour, about which nothing was known at the time except that it was 'peculiar'. It is an irregular galaxy, generally flattened but with a faint filamentary structure extending above and below the disc. The identification was first suggested by C. R. Lynds of the National Radio Astronomy Observatory, Green Bank. He went on to show that the optical spectrum contained very strong emission lines – a sure sign of a high energy concentration.

The Doppler shifts on these emission lines have now revealed the most remarkable feature of M82. It is exploding, the velocities of the outer parts reaching 1 000 km sec^{-1} away from the centre. Extrapolating back into the past, all the material now seen rushing outwards appears to have set off from the centre about 2 million years ago. To add the final touch, the filamentary structure was found to be radiating polarized light – synchrotron radiation!

The Fine Structure of Radio Galaxies

The finest detail to be seen on photographs taken with large optical telescopes under the best atmospheric conditions has an angular size not much less than one second of arc, corresponding to the apparent size of a golf ball seen at a distance of 10 km. A simple reflector radio telescope has a much poorer angular resolution, no better than a few minutes of arc at the shortest wavelength of operation, or even some degrees at the longest wavelength. The fine detail of the emission from radio galaxies can only be resolved by more complicated radio telescopes. There are two types, to be discussed further in Chapter 16. One, the interferometer, consists of two separated radio telescopes joined to the same receiving apparatus: this is used for measuring the angular diameter of an individual radio source. The other, aperture synthesis, is a development of the interferometer. It often uses several radio telescope elements, although it only essentially needs two. These two must, however, be placed at a series of different spacings and orientations, with the effect of building up a receiving aperture equal in width to the largest spacing.

Interferometers can be used for measuring angular diameters as small as one thousandth of a second of arc, but they give very little information beyond the measurement of angular size. Synthesis telescopes are able, in contrast, to produce a complete map of a considerable area of sky to a resolution very nearly as good as that of optical telescopes. The maps represented in Figure 37 were obtained with synthesis telescopes at Cambridge and at Westerbork (Netherlands), which have angular resolutions of the order of 10 seconds of arc (the exact value depends on the

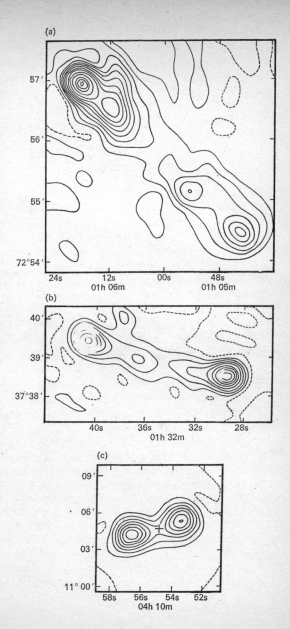

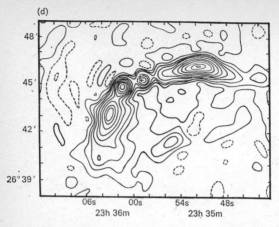

(d)

37. The structure of four radio galaxies. These maps were made by the Cambridge 1-mile telescope at 70-cm wavelength, with added detail obtained at 21-cm wavelength

(a) 3C 33·1   (b) 3C 46
(c) 3C 109   (d) 3C 465

wavelength and on the position of the radio source in the sky).

The very high resolution map in Figure 38, of the radio source 3C 147, is a reconstruction from interferometer observations alone, in which the details may not be exact; there is nevertheless a marked general resemblance to the other galaxies, but on a scale about a hundred times smaller. The most striking feature of all the objects, from the largest to the smallest, is their double nature. Some, like Cygnus A, form a symmetrical pair, while others, like 3C 147 and Centaurus A, are double doubles. Some are much less symmetrical, but even these may be doubles in which one part is stronger than the other.

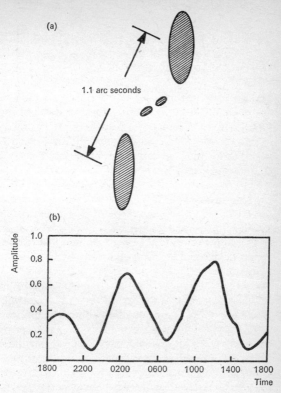

(a)

1.1 arc seconds

(b)

38. The 'double double' radio source 3C 147. (a) The structure of the source, which appears to be a very distant galaxy. (b) Relative amplitudes of interferometer pattern, recorded through one day, from which the structure was deduced (Jodrell Bank/Malvern interferometer)

Interpretation of the Radio Galaxies

The variety of shapes of radio galaxies displayed in these maps is beginning to resolve into a simple classification. At the centre of some radio galaxies, and coincident with the centre of the visible galaxy, there is often a bright spot, so small that it can only be resolved by an interferometer. The main emission usually comes from a pair of bright sources, symmetrically disposed about the galaxy: Cygnus A is the typical double source of this kind. These paired components look like two fragmentary remains of an explosion, moving in opposite directions. There may be more than one pair of sources; there can be a succession of sources forming a series of pairs, or the pairs may themselves show some structure. Finally, there may be a sort of radio halo round the galaxy, which might be regarded as a smoke cloud remaining from past expulsions of component pairs from the galaxy.

The picture indicates an internal explosion, involving a large mass in the central core of the galaxy. From the centre of the explosion two clouds shoot out in opposite directions, containing high-energy electrons and a magnetic field. These clouds cannot exist for long compared with the life of a galaxy, and yet they can still be detected at large distances from their parent galaxies. They must therefore be travelling very fast; calculations indicate velocities several per cent of the velocity of light. No one knows how such a velocity could be imparted to the clouds, or even how they stay so compact, looking like intergalactic missiles. They may in fact be driven by a plasma rocket-mechanism; most theories assume however that they are driven outwards as from a catapult, and given their momentum in one single impulse.

A limit on the velocity which these clouds can have is provided by the observation that most of the pairs have roughly the same radio brightness. Considering that most of the pairs must be moving with one component moving towards us and the other away, we ought to see an effect of special relativity if the velocities are close to the velocity of light, making the closer cloud brighter and the more distant one fainter. This argument fixes an upper limit of about one tenth of the velocity of light; the velocity is therefore known within reasonable limits, even if its origin is not understood.

The containment of the clouds into small packages becomes easier to understand given these high velocities. As the clouds fly outwards they encounter a very thin gas, probably filling the space between the galaxies of a cluster and possibly even the whole of intergalactic space. Figure 39 shows the effect of this inter-galactic medium; there is a ram pressure on the front of the cloud due to its supersonic speed, but the flow pattern round the side of the cloud effectively prevents it from spreading sideways. The shape of the clouds seems to fit in well with this theory.

We return to the question of the energy of the initial explosion in the next chapter, which concerns the early stages of development of radio galaxies.

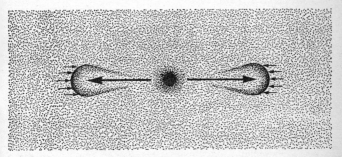

39. Model of a radio galaxy. Gas clouds exploding outwards from either side of a galactic nucleus encounter a very thin intergalactic gas, and are compressed by ram pressure

CHAPTER 11

The Mysterious Quasars

The Unresolved Radio Sources

ALTHOUGH the discrete radio sources in the sky were originally known as 'radio stars', this proved to be a misnomer; identifications came with one galaxy after another, but never with stars. Apart from the identifications, it also appeared that the angular diameters of many radio sources were more consistent with those of galaxies rather than stars. Those with angular diameters less than 1 minute of arc, as yet unresolved, were thought to be rather more distant that the others, and it was confidently expected that they would shortly be resolved along with their larger brothers.

To resolve a small angular diameter a measurement must be made with an interferometer with a large baseline, extending at least some thousands of radio wavelengths. Jodrell Bank, which has specialized in these interferometers, found that the angular diameters of a considerable number of radio sources were unexpectedly hard to resolve, even when the baseline was increased to ten miles or more. Apparently some sources were less than 1 second of arc in diameter. Were these, after all, true radio stars? No galaxy could be seen in their positions, and the first guess at an identification, which was for the radio source 3C 48 (from the third Cambridge catalogue), was for a star-like object of 16th magnitude. The identification, announced by Allan Sandage of Palomar Observatory in December 1960, was based on an accurate position from a new radio observatory at Owens Valley, in California.

Was this identification a chance coincidence in positions, or was the radio source in fact a star within our galaxy? Attempts were made during 1961 to settle this question through studies of the spectrum of the 'star'. These were at first inconclusive, since no one could interpret the rather strange spectrum. A series of photometer measurements showed, however, that the brightness was variable, even on time scales of less than one hour. This can only happen for physically small objects, so that the radio source could not be, for example, a radio galaxy like Cygnus A. The identification with the star seemed to be correct.

Two more similar identifications, of 3C 147 and 3C 273, were made in 1962 and 1963. The determination of the position of 3C 273 was particularly revealing. It was made with the 210-ft Parkes radio telescope, in Australia, using the technique of lunar

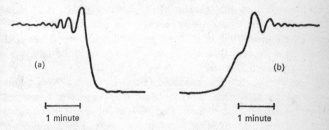

40. Occultation of the quasar 3C 273 by the moon, 5 August 1962, observed at 408 MHz. This famous occultation curve shows the diffraction fringes just before immersion (a) and just after emersion (b). The kink on curve (b) shows that the radio source is double (Parkes 210-ft radio telescope)

occultation. The times of disappearance and reappearance of the radio source as the moon moved across the sky gave a position accurate to about 1 second of arc: further, the occultation took place in two stages (Figure 40), showing that the radio source had two parts. One part, which was the bright unresolved object, coincided precisely with another star-like object, this time of 13th magnitude, so bright that it is easily visible in a 6-in telescope. The other part was more diffuse, and lay about 20 arc seconds from the 'star'.

The excitement of this discovery is reflected in several scientific papers of the time, most of which were made out-of-date before they were even printed. The identifications were already hardly in doubt, but they became quite certain when Maarten Schmidt at Palomar observed that the 3C 273 star had a faint wisp or jet in the position of the second radio component. The two radio sources can be seen on Plate IV, which shows a greatly enlarged image of the 'star' and the jet. The spectra, however, remained a mystery, and the distance of the three objects was still an open question.

The spectrum of a bright object such as 3C 273 is easily measured. It was found to contain emission lines, but at unexpected wavelengths. Schmidt eventually realized that these were the usual lines of hydrogen and magnesium red-shifted by a large factor, impossibly large for a galactic object. The red shift of 0·158 indicated that this star-like object was further away than Cygnus A, even though it appeared to be much brighter. This could only mean that 3C 273 was emitting light several hundred times more strongly than any ordinary galaxy, even though it was obviously much smaller.

It is said that Schmidt reported this awesome discovery to his wife in the following words: 'Something terrible happened today, if what I think is true.' The truth of it became inescapable when the spectrum of 3C 48 showed a red shift of 0·37, and other quasars followed rapidly. The record for distance was soon broken by 3C 147 (red shift 0·57), and then by rapid stages up to a red shift of 2·012 by 3C 9. At this value the wavelength of all the radiation leaving the galaxy is trebled when it is observed from the earth; at the time of emission the galaxy was receding at a velocity of 150 000 miles per second, about four fifths of the speed of light.

The Meaning of the Red Shifts

Optical astronomy depends so much on spectroscopy that a spectrum without an interpretation seems like a shock to the foundations of astrophysics. It was therefore a profound relief to find that the quasar spectra had a simple explanation; but it was nevertheless very disturbing to find that this explanation involved a new type of galaxy, hundreds of times more luminous than any previously seen, but so compact that fluctuations of light output could occur within only a few days. Suppose that the red shift, which made the spectrum look reasonable, was wrongly interpreted? If it did not in fact represent a high velocity of recession, which through Hubble's Law meant a large distance, then the quasars might be close to us, perhaps in our own galaxy. The old arguments about the red shift (Chapter 10) were revived, but with a new twist arising from the theory of superdense stars.

It was now suggested that the red shift was an effect of the very strong gravitational field at the surface of a superdense star. This is a kind of 'tired-light' hypothesis, in which the light quanta lose energy in escaping from a strong gravitational field, and consequently their wavelength is increased as they leave a very condensed massive star. Despite a simple demonstration by G. C. McVittie that this did not fit the evidence, the idea lingered on for several years as a way out of the cosmological problems raised by the distant quasars; it was almost ten years before the attempts to place the quasars nearer home were abandoned.

The magnitude of the red shifts was indeed hard to believe. A red shift z, corresponding to a change of wavelength $\Delta\lambda$ in a wavelength λ, is related to velocity v by the relativistic equation:

$$1 + z = \frac{\lambda + \Delta\lambda}{\lambda} = \sqrt{\frac{1 + v/c}{1 - v/c}}$$

Here c is the velocity of light, which is 300 000 km sec^{-1}. The record value of red shift is now 3·53, for the quasar OQ 172. This corresponds to v/c = 0·907, or v = 272 000 km sec^{-1}. It is not surprising to see some serious doubts about this interpretation: with good evidence, as will be seen from the story of the angular diameters in the next few pages, and from the disturbing fact that some quasars have spectra with absorption lines with a different red shift from that of the emission lines!

The Angular Diameters of Quasars

The early work by H. P. Palmer and his colleagues at Jodrell Bank had already shown that resolution of the angular diameters of quasars was a difficult experimental task. The upper limit of 1 second of arc found for such quasars as 3C 48 was obtained by using an interferometer with a baseline of 60 km and a wavelength of 70 cm; an improved resolution could be obtained either by using a shorter wavelength, which at that time was impossible, or by using a longer baseline, which was difficult but not impossible. Palmer therefore was forced to construct an interferometer with a longer baseline. The difficulty lay in connecting the two elements of the interferometer by radio link, since the length of the baseline now meant that the two ends were beyond the distance of a direct line of sight. The solution was to use a radio relay, situated on top of the Pennine Hills, linking an outstation in Yorkshire with Jodrell Bank.

With this baseline a few more sources were resolved, but the main problem was still untouched. The next attempt used three radio links, running 140 km from Jodrell Bank to Malvern, and a new receiver for a shorter wavelength, 21 cm. Again some sources remained unresolved. The final attempt using 6-cm wavelength put an upper limit of only a few hundredths of a second of arc on the persistent remaining quasars; it also showed that even those with large diameters seemed to be made up of several smaller components, as though they were miniature radio galaxies with double or quadruple structure.

There was now no further possibility of extending the radio links within Britain, and Jodrell Bank gave up the attempt to resolve the smallest of the sources in favour of studies of the structure of the others. The story now moves to a new type of interferometry, which dispensed with radio links and expanded baselines to an intercontinental scale.

The technique of very-long-baseline interferometry, or VLBI, is described in Chapter 16. It involves the simultaneous use of two widely separated radio telescopes, with accurate clocks and tape recorders. A trans-Canadian baseline, between Algonquin and Penticton, was the first to be used, followed by other baselines across continental America, then by rapid stages to a trans-Pacific baseline whose length approached the diameter of the earth. At last the angular diameters were measured; they were a few thousandths of a second of arc. Even at the enormous distances given by the red shifts the actual diameters were only some tens of light-years, and it still seemed possible that there was a structure inside the quasars with smaller dimensions again.

Now came the further excitement which again placed the red-shift interpretation in jeopardy. The VLBI measurements were repeated for several years in succession, and showed that the quasar 3C 273 was expanding. This fitted the theory that radio galaxies grow out of quasars, and added to it the possibility of actually measuring the velocity of expansion. All that was needed was to multiply the distance, obtained from the red shift, by the rate of angular expansion, and the velocity would be found directly. The problem came when the measurements gave a velocity which was several times greater than the velocity of light! The only way out of this uncomfortable conclusion was to reduce the distance scale, which would make nonsense of the red shifts. It was not until 1972 that this problem was resolved by new observations which showed that 3C 273 is not in fact obviously expanding. It is instead a complex source, with several components whose relative brightnesses change within a few weeks. Depending on the way in which these components turn on and off, the source can grow or shrink in apparent size with alarming rapidity. But there is no suggestion now that there is any real expansion at a velocity greater than the speed of light.

One of the experimenters described this behaviour graphically as not the expected steady expansion of a balloon, but the rapidly changing pattern of flashing lights on a Christmas tree. Put in those terms, the incident hardly seemed to be a challenge to the basic astrophysics of the red shift: it was, however, one of the more dramatic moments in the exciting history of the quasars.

What are the Quasars?

Most of the light, and all of the radio emission from quasars is synchrotron radiation. It is not too difficult, given a size, a distance and the strength of the radiation, to follow the calculations previously made for the radio galaxies and find the minimum energy stored in a quasar in the form of high-energy particles and magnetic field. As before (Chapter 10), this can only give the minimum energy, since the calculation must assume the most efficient distribution of energy for the purposes of radiation, and it must ignore any energy stored in other particles, such as protons. The minimum energy for the strongest sources is about 10^{60} ergs; the components which have been ignored could add another factor of a thousand or more to this requirement. Where can 10^{63} ergs come from?

The problem of the energy source already existed for the radio galaxies, such as Cygnus A; there is an obvious possibility that quasars evolve into radio galaxies, so that there is really only one energy problem to be solved. It is, however, an even more difficult problem. There must be a means of generating so much energy not in a whole galaxy, but in a region only a few light-years across, as compared with 50 000 light-years for the diameter of a normal galaxy.

A whole galaxy converted from hydrogen to helium could yield the necessary total of 10^{63} ergs. But to concentrate a whole galaxy into such a tiny volume, and arrange for a completely efficient nuclear burning, is impossible.

These considerations led Fred Hoyle and William Fowler to suggest that the source of energy must be gravitational and not nuclear. Gravitational energy is released by the collapse of a large mass under its own gravitational attraction; for example the energy stored in the rotation of a neutron star is derived from gravitational energy released as it collapses from the size of a normal star. On a galactic scale the condensation of a primeval gas cloud, followed by the formation of stars, all gives a release of gravitational energy. Most of this energy appears in the dynamic motions of the stars within the galaxy, but it is conceivable that a massive gravitational collapse could occur in the dense material at the centre of a galaxy, giving a magnified version of the neutron-star collapse. The supply of gravitational energy from such a collapse could be about 100 times larger than the nuclear energy available from the same mass of material.

If this is correct, a collapse of something like 10^9 solar masses is required to explain the energy of quasars and radio galaxies. This is only 1 per cent of the total mass of a typical galaxy; such concentrations of mass are not unusual at the centres of elliptical galaxies. The detailed process of converting the energy of collapse into magnetic fields and the electric fields needed to accelerate electrons to cosmic-ray energies is not, however, understood.

We now leave to one side the internal problems of the quasars.

Accepting that they represent a special kind of galaxy, or a special phase in the development of most galaxies, we can now use them as signposts on a journey of cosmological exploration into the distant and most ancient parts of the universe.

CHAPTER 12

Radio Cosmology

THERE is a conventional view of science in which progress is made by following a sequence of hypothesis and experimentation, known as 'the scientific method'. It is doubtful whether such a set pattern actually exists in any science, but it is certain that it does not apply in cosmology. The trouble with cosmology is that there is only one universe, so that there are no fundamental laws of the behaviour of universes in general to be discovered. Furthermore, there are no experiments to be performed on the universe, as there are in laboratory physics. We can only observe, and describe the universe surrounding us.

The description of the universe is on two entirely different scales, which may be called the general and the particular. We have so far in this book dealt with various aspects of the particular, where we are concerned with diverse discrete objects such as the sun, the Milky Way, and the distant galaxies. The general, large-scale description ignores all of these discrete objects as irregularities in a smooth background, whose properties are averages of quantities such as density, velocity, temperature, age, and a distance scale. The outstanding feature of this general description is that the universe is the same in any direction in which we look. We seem to be in the centre of symmetry, at the hub of the universe. Are we really at the centre, or is this only an illusion?

The Einstein–de Sitter Universe

Any description of the universe must take into account the further problem that light, which brings us our information about the distant parts of the universe, travels with a finite speed. Our description therefore is of distant parts as they were a long time ago, while the view of the nearer parts is more modern. Einstein formulated a mathematical description of the universe in which his equations described a static situation, with no expansion in the distant parts. The key factor in arranging for a static universe is its mean density, since gravitation is always tending to shrink the distance scale, or at least to slow down any velocity of expansion. Later, de Sitter showed that Einstein's equations need not apply only to a static universe, but with a reasonable mean density the universe could be expanding with a velocity proportional to distance from the observer.

These mathematical models provided little more than a possible theoretical framework for the interpretation of actual observations. At the time of Einstein and de Sitter there was only one real observational fact, remarked on by Olbers as early as 1829. Olbers pointed out that the darkness of the night sky could not be accounted for in any model universe which was infinite in extent, and whose properties did not change with distance.

Olber's Paradox, as it is usually called, is simply that any line of sight into an infinite universe will eventually meet a bright star, so that the sky should be as bright as the sun in all directions. It can be resolved either by supposing that the density of stars in the universe falls with distance, or that the universe is expanding, so that distant objects are moving away from us and appear to be dimmed by the Doppler effect. The first approach places us in the centre of the universe, a position which most theorists prefer to avoid. The second does not; in a uniform expansion an observer at any point sees all others receding from him isotropically.

The Einstein–de Sitter theory therefore prepared the way for acceptance of Hubble's observation that the red shifts of galaxies increased fairly uniformly with their distances. There was, however, a range of models which fitted Hubble's observation. The main variation concerned the mean density of the universe, which determined whether or not the expansion would slow down under gravitational forces, including the possibility that the expansion might eventually reverse, giving the oscillatory universe of Figure 41. A test of this seems to involve an impossible look either into the past or the future, to see whether the expansion remains constant. Looking into the past is, however, not impossible, as it is synonymous with looking into the distance – 'far away is long ago', as the cosmologists say.

At this point optical astronomy gave no further answers to cosmological questions, since the galaxies that could be observed optically were not far enough away for any such distance effects to be appreciable. The best measure of distance on the cosmological scale is the red shift itself. Optical galaxies rarely reached a red shift of over 0·2, while a critical test of cosmological theories really needed values of over 1·0 and preferably as large as 2. Even before the discovery of the quasars, some of which have red shifts well over 2, radio astronomy was able to provide the first real evidence on the structure of really distant parts of

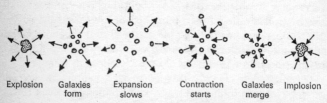

| Explosion | Galaxies form | Expansion slows | Contraction starts | Galaxies merge | Implosion |

41. The evolution of an oscillating universe. After the initial explosion (the 'big bang') the expansion slows down progressively. When the mean density of all matter in the universe is approximately 10 atoms per cubic metre, the expansion halts and the universe contracts again to a single body. A further explosion then starts the cycle again

the universe, where conditions are sufficiently different for cosmological tests to be applicable. The new evidence came just at the right time to test a new and exciting theory, known as the 'steady-state' theory.

The Steady-State Theory

In 1948 Hermann Bondi, Thomas Gold, and Fred Hoyle proposed the simplest model universe that had yet been conceived. Starting from the basic postulation that we, as observers, are not privileged in any way to obtain a special view of the universe, so that the universe should appear on average the same from the point of view of an observer at any location in the universe and at any time in its history, they constructed a self-renewing universe known as the 'steady-state' universe. In this the outward flow of matter, due to the observed expansion, was counterbalanced by a continuous creation, atom by atom, at a rate too small to be detectable in a laboratory but sufficiently large to account for the creation of the material for new galaxies to replace the old, receding galaxies.

The steady-state universe was appealingly simple. It was also precisely specified, and subject to test by comparison with observation. In contrast to evolutionary theories of the universe, which allow a sequence of events, such as an explosive expansion from an initial concentration, or a slowing and reversal of the expansion, the steady-state universe had no different behaviour in the past, no history that could appear in observations of the distant galaxies, where 'far away is long ago'. The test of the steady state came when radio astronomy gave the first look far enough into the distance to see real differences between the past and the present.

The Surveys of Radio Sources

The identifications of the first few radio sources provided a confusing mixture of objects within and outside our galaxy. Attempts to survey the sky for more sources were made both in Cambridge and in Sydney, leading at first only to controversy concerning the reality of the results, but eventually to a proof that there were very many extragalactic sources, and that they were far enough away to be profoundly important in cosmological theory.

The first such survey in Cambridge detected and located fifty radio sources. The second, known as 2C, produced a catalogue of nearly 2 000. Both surveys used interferometer radio telescopes, which were designed to be sensitive to radio sources with small angular diameters. In Sydney a survey using a different type of radio telescope (the Mills Cross: see Chapter 16) was at this time giving a rather different result; it seemed to be showing that there were many sources with rather large angular diameters, which would not be properly recorded by the Cambridge instruments. Although this subsequently turned out to be not the case, it formed the basis of strong Australian criticism of the Cambridge observations. This criticism found some material and disturbing support when surveys of the same part of the sky were compared. Hardly any of the individual sources corresponded; disagreement was almost complete. The reason turned out to be more serious than the possible misrepresentation of the large-diameter sources.

The trouble largely lay within the Cambridge survey. An early warning of trouble had already appeared in the paper in which Ryle and Smith described the discovery of Cassiopeia A. Here they also mentioned a strong radio source in Ursa Major, giving a position and intensity. Not long after publication they found that this radio source did not exist at all, but that the radiation from Cassiopeia A was being picked up in a 'side-lobe' of the interferometer aerial so that it seemed to come from a quite different direction. This particular problem was, however, understood and dealt with before the 2C survey was made; the real problem, known as 'confusion', was not revealed until the results of the 2C survey were compared with the survey by Mills in Sydney.

'Confusion' arises from the limited angular resolving power of radio telescopes, combined with the way in which their signals are recorded. In an optical photograph of the sky each star is clearly distinguished, or 'resolved', from its neighbours. The output of a radio telescope may at one moment derive from a large area of sky, such as the area covered by a photographic plate, but because of the poor angular resolving power of the telescope everything in that area combines to produce one single output. If there is one source in the area brighter than the others, that source will dominate the receiver output; nevertheless its record will be confused by the combined effect of the others, and interpretation of the record can give errors both in brightness and in position of the source. Worse, the combined effect of the small sources may give the appearance of a single bright source, and a fictitious source may then find its way into a catalogue. The problem has in fact been overcome in the modern aperture-synthesis telescopes, but it was for many years a severe limitation in all radio surveys.

The confusion problem in radio-source surveys was for many years the cause of a fierce argument between the observers at

Sydney and at Cambridge, with the theoretical cosmologists cheering on the contestants from the sidelines. When the 2C was found to be seriously affected by confusion the protagonists of steady-state cosmology cheered, since the survey had seemed to show that the universe was very different from steady state. The reply from Cambridge was that the unexpectedly large amount of confusion in the survey showed that there was an unexpectedly large number of distant radio sources; this statistical answer was not easily accepted, even though it subsequently proved to be correct. It was not until two further major surveys, the 3C and 4C, had been made at Cambridge, that the cosmological significance of the radio-source counts was accepted internationally.

The Statistics of the Radio Sources

The expanding universe is often compared with an infinitely large plum pudding, growing uniformly as it cooks. Each plum represents a galaxy, and each plum is surrounded by plums retreating steadily from it, uniformly in all directions. Each plum moreover will think itself the centre of the pudding, as not only are the motions of the other plums radically outwards from itself, but their velocities increase in proportion to their distances. It is only if the pudding has an observable edge that any plum could in fact have the right to claim a position at the centre. This model serves to remind us that distant nebulae are observed distributed uniformly over the sky, and that the number seen out to a certain distance increases simply according to the volume of space out to that distance. Moreover the distance also determines the intensity, S, of radio waves from the nebula. But we have seen that in the most distant parts of the universe we may find it worth while to look for differences in the age, and therefore the state of expansion, of the plum pudding, and the number of nebulae will not increase by such a simple law. This simple Newtonian law states that the number N is proportional to the minus three-halves power of S, so that a logarithmic graph of N against S would have a slope of minus 1·5. In an evolutionary universe the law would not be so simple, and the graph would not follow the same straight line.

The first opportunity to plot this graph came with the results of the 2C survey, whose 2 000 radio sources appeared to be enough for a significant test of the law. Ryle announced the result in the Halley lecture in Oxford in 1955, when he showed the graph of

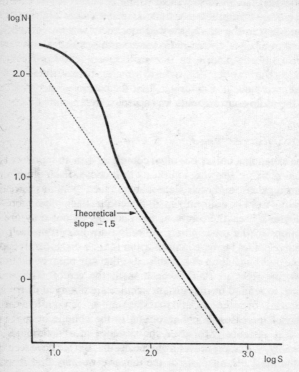

42. Radio source statistics. Plot of Number N against intensity S showing an excess of weak sources

Figure 42. This graph was obtained from radio sources covering most of the sky visible from Cambridge, but it was also possible to show that the same graph could be obtained from all separate parts of that sky except near the Milky Way, where galactic sources are found. The graph showed not a slope of −1·5 as expected, but a slope of about −3; the top part of the graph where the line tails over can be ignored, as this is the region where the radio sources are too weak to be detected individually. A slope of −3 indicates that the radio population increases with increasing distance, that is to say that we are in a kind of spherical hole where there is a

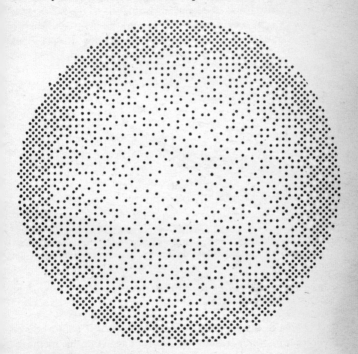

43. Distribution of the galaxies in an evolutionary universe

comparative dearth of radio objects. How big the hole is needs further argument, but Ryle showed good reasons for suggesting that it is at least 300 million light-years across, implying that the increasing number of radio sources seen at low intensities are all at distances comparable with limits of the observable universe. This would give a picture of the universe like Figure 43. If on such a scale the universe is found to be so non-uniform, the steady-state hypothesis of the universe must immediately be abandoned, and the explosion theory is given strong support.

Unfortunately it did not turn out as simply as that. The Mills survey in Australia did not confirm the shape of the curve, and moreover showed that the 2C catalogue was more in error than anyone had suspected. A complete re-analysis of the 2C records by a new statistical method was made, and this showed a smaller slope in the graph of Figure 42; even though this still provided

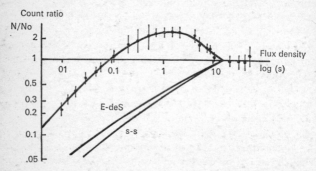

44. The counts of weak radio sources. For a simple Newtonian universe the counts would lie on the line $N/N_0 = 1$, representing the simple 'three-halves' power law. For Einstein–de Sitter and steady-state universes the expected curves are below the line. The actual counts show first an increase as the strength falls from 10 units to 1 unit, then a decrease for the weakest sources. The vertical 'error bars' through the points are an indication of their accuracy (after M. Ryle)

clear evidence against the steady-state theory it could not be accepted until similar results had been obtained by other methods. Final agreement between the Australian observers and the Cambridge group was only reached in 1964, when the latest surveys settled on a slope of -1.8 for the initial part of the curve. Figure 44 shows the present state of the curve, in the form of differences from the simple -1.5-power law. The initial rise above the -1.5-power law persists down to one flux unit on the intensity scale; below this the numbers of sources fall sharply. At the peak there are almost three times as many sources as are predicted by the simple Newtonian model, while at the lowest intensities there are fewer than one tenth of the expected sources.

These statistics might, of course, refer to a very local part of the universe, where we are located by accident in the centre of a spherical bubble or shell of radio sources, with only a few sources inside or outside the shell. We have as evidence against this improbable situation a radio version of the original Olber's Paradox. The radio sources which can be observed individually in these counts already account for almost half of the total radio brightness of the sky, so that there cannot be other bubbles of the same kind, or any other formations or distributions of radio sources in the universe. The radio surveys penetrate so far into the universe that they are able to detect as individuals a high proportion of all the radio galaxies and quasars which exist in the observable universe.

Here is positive proof that the universe is evolving. The choice among the Einstein–de Sitter and other models is still to be made: the only clear result is the disproof of the steady-state universe. The possibilities which remain are usually known as big-bang universes, in which the galaxies rushing apart so dramatically at this time are flying apart from a central explosion which occurred 10 000 million years ago.

The Big Bang and the Relict Radiation

The next stage in unravelling the history of the universe is to explain the existence of galaxies. The physical conditions in the early explosive stages of the expanding universe are simple: all material was in the form of a uniform hot soup, filled both with matter and radiation. The proportions of different kinds of atom, and the density of the radiation, depended on the temperature, which fell uniformly as the expansion proceeded. At a certain stage of expansion the ingredients of the soup fell apart, so that the temperature fell without the densities of matter and radiation continuing to adjust to equilibrium values. The remaining radiation from this stage is known as the 'relict radiation', while the dispersing atomic material is available for the formation of galaxies and stars.

No one has yet produced a complete theory for the next stage of evolution, in which the uniform distribution of matter changes into the material condensations of the galaxies, just as some milky varieties of soup tend to curdle. It apparently happens fairly suddenly, at an epoch just before that of the maximum concentration of radio sources. Most of the new galaxies are likely to develop violent instabilities near their centres, the most violent of which are now visible as quasars. There may be repeated explosions, leading to the expulsion of blobs of hot ionized gas, or hot plasma, in two opposite directions. These then become the components of the radio galaxies.

This sketchy outline of the first stages of the history of the universe may seem too fanciful and too remote for any real observational test to be relevant. There is, however, a direct observation of what is believed to be the relict radiation, dating back to before the formation of galaxies. It has been described by the cosmologist Dennis Sciama as 'perhaps the most magical of all astronomical discoveries'.

The expected intensity of the relict radiation was calculated in 1965 by Robert Dicke of Princeton University. The intensity, and the spectrum, was that of thermal radiation from the whole sky, with a temperature of a few degrees absolute, and certainly less than 40 K. Dicke realized that this very weak radiation could be detected at short radio wavelengths, and he started an experiment to search for it at 3-cm wavelength. Soon after, he heard that two radio engineers at the Bell Telephone Laboratories, Arno Penzias and Robert Wilson, had been making very accurate measurements of the radiation from the sky at 7-cm wavelength for an experimental satellite-communication system. They were puzzled by a signal which came from all directions in space, very much stronger than the sum of the signals from the discrete radio sources, and with no obvious origin. Its intensity corresponded to a temperature of 3 K.

Very few of the exciting discoveries of radio astronomy have been made as a result of theoretical predictions. The discovery of the 3 K background radiation, which must rank as one of the greatest astronomical discoveries of all time, was made accidentally and without any knowledge of the prediction. The remarkable fact of its existence is matched by the accuracy of the predicted intensity. Dicke had calculated the equivalent temperature of the radiation by assuming that radiation and matter became 'decoupled' only 100 seconds after the big bang, when the temperature was 10^9 degrees absolute. Thenceforward the temperature decayed, falling with the square root of the time after the big bang. The 3 K radiation is indeed a relict radiation: anyone who detects it in a short-wavelength radio telescope ought to pause and contemplate the ease with which he can look at the traces of the most ancient event imaginable.

Angular Diameters – the Curvature of Space

There is one further aspect of cosmology in which radio astronomy offers the chance of observational test. This was suggested by Hoyle before the source counts had given the first definite results, but it is still unrealized despite its apparent simplicity. It is essentially a measurement of the strange geometrical properties of space on a large scale.

On a naïve Newtonian view the small-scale geometry formulated by Euclid can be extended indefinitely to cover all of space. Relativity physics shows us that in a large-scale universe containing matter, Euclidean geometry cannot be correct: it is not even a simple matter to define a straight line, and no longer true that the angles of a triangle add up to 180°. These deviations from Euclid are attributed to a 'curvature' of space, since they are very similar to the differences between the geometries of plane and spherical surfaces. The amount of curvature in space is a parameter distinguishing between the various model universes such as those of Einstein and de Sitter.

A measurement of the curvature of space would be provided by a measurement of the changing angular diameter of an object placed at successively greater distances from the observer. On most cosmological models the angular diameter is expected to fall steadily with distance, at first inversely with distance, but eventually more slowly. In some cosmologies the angular diameter then reaches a minimum, and surprisingly then increases. The distance when the angular diameter reaches a minimum is approximately that distance where the red shift has a value of one.

There are now many radio sources known with red shifts of unity and greater. If they all had the same linear size and were at different distances, then a measurement of their angular sizes would provide the required test of the curvature of space. Hoyle suggested making the test on the basis of the identified radio galaxies Cygnus A and 3C 295, which seemed to be the first of such a series. The steady-state cosmology predicted that the angular diameter of radio galaxies of this kind would fall with increasing distance to a minimum of 1 second of arc, and not below. The test failed with the discovery of the quasars, some of which have angular diameters only 0·001 seconds of arc. This small angular diameter was nothing to do with the curvature of space: it just showed that there were small radio sources as well as large ones. Radio galaxies and quasars do not form clearly distinguishable classes of objects, so that there is at present no single class with well-defined properties that can be used for this geometrical test. Radio astronomy has given us a surprising access to the history of the universe, but not as yet to its geometry.

CHAPTER 13

The Moon

IN the days when astronomy used to be a purely geometrical science the moon was regarded as by far the best astronomical time-keeper, moving round the earth in a precisely determined orbit. The orbit was complicated, but the motion of the moon was predictable so exactly that the time at which it cut off the light from the fixed stars could be used for setting the most accurate clocks. The atomic clocks have changed our attitude, and we now determine the finer points of the moon's motion, along with the irregularities in the earth's rotation, by using ammonia or caesium oscillators to mark out the even passage of time.

The motion of the moon is still its best known and best understood characteristic. Detailed and beautiful photographs of the surface are now available from rocket explorations, manned and unmanned. Samples of surface materials are being analysed in laboratories all over the earth, and there has even been seismic exploration of the interior of the moon. But it is still true to say that much of the nature of the moon, and of its history, still eludes us. In many ways it is a more complicated body to understand than a star, where simple physical principles enable us to observe the light emerging from the surface and to deduce the conditions in the interior. We know almost nothing about the interior of the moon, and we have only the most speculative theories to account for its formation. But it must surely be one of the most interesting and important celestial bodies, affecting everyday life far more than any distant star or nebula. We could perhaps manage without moonlight, although this would seriously inconvenience poets, madmen, and lovers, but losing the tides with which the moon enriches our shores would be too much for a seafaring and holiday-making nation.

The geometrical part of the description of the moon which has been arrived at from conventional astronomy is quite precise. The moon is travelling in a nearly circular orbit round the earth, at an average distance of 464 000 km (240 000 miles). A good motor car might cover this distance in its lifetime; modern rockets manage it in three or four days. From the distance and from the angular size of the moon's disc, which is about $\frac{1}{2}°$, one finds a diameter of 3 476 km, about a quarter of that of the earth. The moon rotates once in $29\frac{1}{2}$ days, which is the same as the orbital period. The rotation presents to the sun a changing aspect of the moon, so that sunlight follows a cycle of $29\frac{1}{2}$ days on the moon's surface; the aspect presented to the earth does not change, apart from a small oscillation known as 'libration' (see page 184). This locking of the moon's attitude towards the earth is a result of gravitational forces, which pull on one face more strongly than the other and prevent rotation.

Space exploration has transformed our knowledge of the moon so completely that the contribution of radio astronomy may seem to be minor or historical only. There is, however, some convenience in studying the moon from an earth-bound observatory, and a great advantage in being able to study the whole surface rather than a small sample close to a lunar landing point. Radio techniques are, of course, only supplementary to the vast body of existing optical work, but there are some aspects where the use of long wavelengths has a particular significance. These concern the possible existence of a tenuous atmosphere round the moon, and measurements of temperature of the surface. Radar observations also have a special significance, since they give both a description of surface topography and an indication of the material at some distance below the surface.

The Moon's Atmosphere

We are all familiar with the necessity for the astronauts to carry with them their own oxygen supplies. It is nevertheless hard to appreciate how good is the vacuum in which they must operate, and it is a matter of considerable interest to try to measure the actual density of any traces of atmosphere which the moon may have. It is easy to see from earth that there is very little atmosphere – one need only look at the edge of the moon through a low-powered telescope and observe the stars as they pass behind and are occulted by the moon's disc. If there were an atmosphere like our own, the stars would become fuzzy and their apparent positions would change as a result of the passage of their light rays close to the surface. No such effect is observed, and the most careful measurements at the moment of occultation show that there is no atmosphere greater than 1 000 millionth (10^{-9}) of the earth's atmosphere. Perhaps one should be content with this limit, because it already corresponds to as good a vacuum pressure as we can obtain in a physics laboratory with common techniques. But there are particular theoretical reasons for wanting to reduce the limit still further.

The gravity on the surface of the moon exerts a force only one twenty-seventh of that of gravity on the earth's surface, and any atmosphere on the moon must be maintained there by this force alone. Working against this force is the thermal velocity of gas molecules; for example in air at the temperature of the moon's surface the average velocity is about 0.3 km sec^{-1}, while if the molecules attain a velocity of 2 km sec^{-1} they can escape altogether from the moon. There is a wide spread of velocities about the average, so that a considerable proportion of air molecules would be leaking away, and any atmosphere would soon be lost entirely. Heavier molecules, such as those of the gases xenon and krypton, move more slowly, and would more easily form a lunar atmosphere. Curiosity drives us to look for an atmosphere of these rare gases, although something quite different might turn up if the limits of detection could be improved by a factor of a hundred or so. This improvement is made in the observation of the occultation of a cosmic radio source as it is covered by the moon.

This radio experiment is very much the same as the optical observation of a star just at the point of occultation, but the sensitivity to a lunar atmosphere is tremendously enhanced by the ionized state in which any lunar atmosphere would be found. This ionized atmosphere would have a potent influence on long-wavelength radio waves from a radio source passing near to the disc of the moon, and would show its presence by cutting off the radio waves earlier than the expected moment of occultation. On 24 January 1965 the moon passed across the line joining Cambridge to the Crab nebula, which is a strong radio source, and the radio waves were observed to be occulted for 59·6 minutes. According to calculations, the disc of the moon took only 59·2 minutes to pass over the nebula, and the difference of 0·4 minutes, although not large enough to be a certain and accurately measured difference, looked very much like the effect of a lunar atmosphere.

Interpretation of this result by B. Elsmore at Cambridge shows that the vacuum on the moon is in fact better than had been expected. If the time difference were real, this indicated an atmosphere 5 000 times less than the previous maximum estimate, that is to say two parts in 10 million million of the earth's atmosphere. This is indeed a small amount of gas: if the whole of the lunar atmosphere were collected together and compressed to our normal atmospheric pressure there would be just about enough to fill St Paul's Cathedral. Regrettably, we do not yet know what gas it is, but it would be unlikely to be found to be a breathable atmosphere even if it could be collected together.

One consequence of this radio observation is that we now know that the edge of the moon is nearly as precisely defined for radio waves as it is for light, so that occultation of radio sources can be used as a precise means of finding their positions and for exploring the distribution of radio brightness across them. At short wavelengths any ionized atmosphere has no effect, and the edge of the moon cuts across the source like a knife with an edge sharper than 1 second of arc. A radio lunar occultation of the source 3C 273 played a most important part in the discovery of quasars, as it gave a position accurate enough for an identification and it showed that part of the source was less than 1 second of arc in diameter.

The Temperature of the Moon

There are many ways of measuring the temperature of a body which is accessible to laboratory experiments, but for an inaccessible body like the moon it is only possible to measure its electromagnetic radiation, for example in the infra-red region of the spectrum, and deduce from this its temperature by using well established laws of thermal radiation. At first sight this appears impossible for the moon, for it is obvious that the light which comes from the full moon is reflected sunlight and has nothing to do with the moon's temperature, while even the faint light from the disc of the new moon comes from sunlight reflected off the earth. Visible light, however, is in a part of the spectrum where the sun radiates very strongly, and the moon very weakly. The position changes rapidly as longer parts of the spectrum are considered, and infra-red observations already give the possibility of thermometry at a distance. Radio waves are almost entirely free from trouble from solar radiation; the appearance of the moon for our new radio eyes changes not according to the illumination at any part of the lunar cycle, but according to the surface temperature only.

The first radio measurements of the moon's temperature were made in 1948 by Piddington and Minnett of the Radiophysics Laboratory in Sydney. They used a wavelength of 1·5 cm. Subsequent measurements made in Russia and in America have used wavelengths of only 8 mm. A radio telescope for such short wavelengths is a very compact instrument, with a parabolic reflector about 1 m across and with a waveguide measuring about 3 × 4 mm inside picking up the radio waves at the focus. The beamwidth of one of the best instruments, used by Gibson in Washington, was only 0·2 degrees, which is considerably less than the angular diameter of the moon, so that it was possible to explore the surface from one side to the other and to look for differences in temperature. The main experiment, however, was to watch the centre of the moon's disc throughout a lunar month, which gave a result like the other experiments following the graph of Figure 45. The graph shows variations in the moon's surface temperature with its changing phases, of about 30 K above and below the level of 190 K, which is roughly the average surface temperature.

During Gibson's experiment, there were two occultations of the moon, when the light and heat from the sun were cut off by the earth's shadow passing across the moon. Cold as it is, one might expect the moon to become suddenly colder when the sun is covered in this way, but no change could in fact be found during the whole occultation. This agrees with the rather small variations of temperature found between the full and new moon, and suggests that the moon is a well-insulated body, like a well-lagged water-tank which keeps its temperature regardless of changes in the outside air.

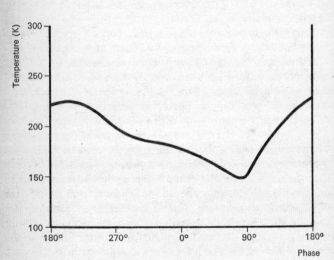

45. The radio temperature of the moon through one lunar month

One factor must be remembered in interpreting these results. Light from the moon comes from the surface, that is from the top few layers of molecules in the surface material, be it dust or rock. Radio waves come from a thicker layer, which may be more than a whole wavelength thick depending on the material. The intensity of the radio waves is therefore determined by the

182

average temperature of a surface layer at least several centimetres thick; for the shortest wavelength used so far, 8 mm, this thickness is about 6 cm or just over 2 in. When the sun's radiation is cut off the actual surface temperature must fall considerably, so that the constancy of the radio temperature tells us that the surface layer is a very good thermal insulator, preventing rapid changes of temperature from penetrating as far down even as one inch. Solid rock would not behave in this way. On domestic water tanks and between the rafters of roofs various kinds of granular substances are used to provide efficient heat blankets in our houses; if the air could be removed from the interstitial air spaces they would be very much more efficient. The insulating properties of the moon's surface therefore led radio astronomers to suggest that there was a thick layer of dust over most of the surface. The correctness of this view was demonstrated immediately the first astronaut, Armstrong, set foot on the moon.

It is an amusing thought that we can in principle explore the emperature of this layer at various depths down to several metres by measuring with progressively longer radio wavelengths. The longest wavelength for which this has been achieved is 70 cm. At this long wavelength there is practically no variation of temperature through the lunar cycle, and very little variation over the disc of the moon. The measurements are unfortunately not accurate enough to give an indication whether the inside of the moon is hot or cold; the accurate measurements of temperature gradient are best carried out by the astronauts. As the archaeologists say, there is in the end no substitute for digging.

Radar Astronomy of the Moon

The achievement of first detecting radar pulses reflected from the moon stands to the credit of a Hungarian, Z. Bay. It need not be emphasized that by the time a radar pulse has reached the moon, part of it scattered and part absorbed, and some of the scattered part has returned to earth, the power in the pulse that enters the receiving aerial is very small indeed and hard to detect against the random signals to be found in any receiver. Bay's solution to this problem was original and noteworthy. He decided to look only for pulses returning exactly 2·5 seconds after transmission, and, to sort these out from the receiver noise, he averaged the signals given by the whole instrument at several separate time intervals after transmission, and looked for an excess of signal at the 2·5-second mark. A long time was necessary for this averaging to be completed, and the signals were therefore added up for over half an hour in water voltameter cells. The detection of moon echoes was demonstrated by the relative amounts of gas liberated in the different cells. Chemistry does not appeal to most electronics experts, so it is not surprising to hear that the method has not been copied in other moon radar equipments.

An American Army radar made contact with the moon at very nearly the same time as Bay. The method was more conventional, and the difficulties of detection were overcome by the use of a larger aerial and greater transmitter power. These two experiments were the forerunners of a series conducted in the three observatories of Sydney, Jodrell Bank, and Washington, which were primarily responsible for the more precise investigations of moon radar echoes, from which came some remarkable new ideas on the form of the moon's surface.

The outstanding characteristic of the echoes received from the moon is their variability. They fade in the same way as long-distance shortwave transmissions, and their amplitude can fall to zero instantaneously and rise several times above the average immediately afterwards. This behaviour was first investigated by Rayleigh as a part of probability theory, and it is often called Rayleigh fading. It is characteristic of simultaneous transmission over several paths together, combining in varying relationships to give additive or destructive interference according to the different lengths of the paths. In a notable Australian experiment, in which the broadcasting station 'Radio Australia' was used as the transmitter, groups of three pulses were sent up to the moon; the returned signal was fading so fast that quite often out of a group of three only two would be heard, one having apparently got lost on its long journey. The speed of fading is proportional to the radio frequency being used and also varies according to the peculiar motion of the moon known as 'libration'.

Libration is an oscillation of the moon, a rocking of the moon's face from side to side so that the visible part of the moon includes positions at either side which periodically come into view. A full cycle of libration brings into view about 58 per cent of the moon's surface. During the cycle there are times when the surface is moving at about 4 kilometres per hour across our line of sight to the centre of the moon, and others when it is momentarily stationary. The speed of fading is greatest when the libration velocity is greatest, confirming the suggestion that the fading is caused by the relative motion of different parts of the surface responsible for the multipath transmission. At Jodrell Bank it was shown that the speed of fading could be used to find how much of the moon's surface was effective in returning an echo; from this work J. Evans made the first suggestion that the moon was not behaving as might have been expected. It seemed to be rather shiny, more like a mirror than a rough-textured sphere with a rugged surface covered with mountains and rocks. A more direct approach was possible when the Naval Research Laboratory published results of radar work using very short pulses. A radar pulse takes longer to return from the limb of the moon than it does from the centre, by 11·5 milliseconds. The first experiments had used pulses which lasted much longer than this, and it was impossible to sort out the part of the echo which came from different parts of the moon's surface. The new experiments used pulses only 10 microseconds long, giving a resolution of more than one thousandth part of the depth of the moon. Individual echoes still showed fading, but an average echo showed that the only parts of the moon which contributed significantly to the echo were those lying within 15 km of range, corresponding to about 10 microseconds travelling time. Figure 46 shows how little of the moon's surface can be reflecting the radar pulses back in the direction of the earth. An area about 300 km across contributes over half of the total echo.

Being wise after the event, it is easy to point out that this behaviour is exactly what would be expected from a moon whose surface is covered with sand or dust. The surface will be smooth but gently undulating, and all but the sharpest mountain peaks will have their sides smoothed over by gentle slopes of powder. The absence of any appreciable atmosphere on the moon means that 'weather' is an unknown quantity there. No winds or rains disturb the quiet deserts of dust. To visitors from earth the land-scape is forbiddingly bleak, silent, and cheerless. For radio waves a smooth dusty surface is as good as a mirror provided that there are no rocks or other irregularities more than a fraction of a wavelength across; the radar reflections show that roughness on this scale is quite rare. The apparent smoothness also suggests that over large areas of the moon's surface there are no slopes

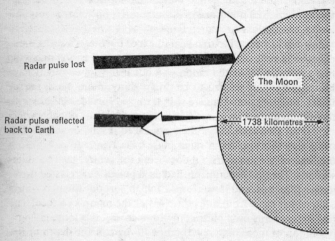

Radar pulse lost

The Moon

Radar pulse reflected back to Earth

←1738 kilometres→

46. The reflection of a radar pulse by the moon

steeper than about 7°. Admittedly, terrestrial slopes are not often much greater than 7° if averaged over several kilometres, but inside shorter distances any sort of gradient can be found. To make the earth look like the moon it must be desiccated, sand-papered, and sterilized.

Radar Aperture Synthesis

In previous chapters we have seen how the radio pictures of radio galaxies and quasars have been improved by the technique of aperture synthesis, in which the angular resolution of a very large aperture is obtained by the use of an interferometer system. The same development is almost impossibly complicated for a radar, since the whole system must be used as a pulse transmitter and then be switched to become a receiver. A different approach has therefore been adopted, called radar aperture synthesis, which needs only one antenna. The process of synthesis now depends on a relative movement of the radar station and its target, so that the target provides a varying aspect. This is provided by the libration of the moon: it is also possible to some extent for planetary radar through the motion of the earth in its orbit.

The possibility of radar aperture synthesis is inherent in the fading of echoes described in the last section. The positions of the discrete features which cause the fading determine the rate at which the echoes fade. Full information on their positions can only be obtained by measuring the amplitude and phase of the echo signal over a long observing period, usually more than 12 hours, chosen as a period during which the aspect of the moon changes as rapidly as possible. The echo information is then put together in a way similar to the ordinary aperture synthesis, and a map of the moon is obtained. What does this map tell us?

The map is a radar map, not a map of emission. It measures the reflection properties of the surface, which are determined by the material of the surface and its physical state but not by its temperature. The details of the surface cannot be made out as clearly as in an ordinary photograph of the moon, particularly for radar mapping at long radio wavelengths. Figure 47 shows a radar map obtained at the long wavelength of 70 cm. This map was made at Jodrell Bank in 1968 by the originators of the radar-aperture-synthesis technique, J. H. Thomson and J. E. B. Ponsonby. It shows different reflectivities for the mountains and the flat areas of the moon (the 'highlands' and the 'maria'). The outstanding feature in the southern half of this map is the large crater known as 'Tycho'; this crater is a very good radar reflector.

On shorter wavelengths the detail of quite small craters can be seen very clearly. Their greater reflectivity may be an effect of their rocky surfaces, as contrasted with the dusty surfaces of the maria. However, the radio waves do penetrate to a depth of several wavelengths; they are therefore particularly useful in

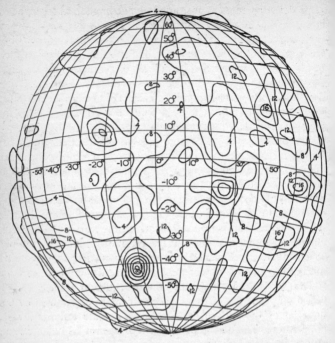

47. Radar map of the moon using aperture-synthesis radar at 70-cm wavelength (Jodrell Bank)

exploring conditions below the surface at many locations away from those where the astronauts are able to take direct samples.

The Moon as a Relay Station

The use of artificial satellites for long-distance communications, and especially for television relays, has become a commonplace. The first satellite communication system, however, used our natural satellite, the moon. In 1954 the Naval Research Laboratory, Washington, showed that speech could be relayed via the moon, by picking up the reflected signals from an ordinary transmitter working on a frequency of 220 MHz.

It is recorded that Alexander Bell, on making the first telephone link between two towns, opened the circuit with the words, 'What hath God wrought'! No record exists of the words used on the first speech circuit via the moon, but it must be admitted that this was the lesser revolution in communication history. Before the discovery of the shininess of the moon's surface it was supposed that the lengthening of all reflections from the moon by 11·6 milliseconds would so distort any speech that only very slow transmission of information would be possible. Now that the lengthening is known to be only 100 microseconds it is apparent that a communication circuit is available which is capable of transmitting normal speech and possibly even music with very little distortion. Modulation frequencies of up to 10 KHz, higher than those passed in a normal radio receiver, could be transmitted without attenuation. There are, however, considerable difficulties.

The first difficulty in a moon relay circuit is to obtain enough power for communication to be established at all. This certainly means highly directive aerials, high-power transmitters, and sensitive receivers, all of which are possible to make, even if they may be difficult or expensive. Secondly, the signal will fade so deeply that it will be necessary to provide at least two circuits working simultaneously to ensure that a signal can be received at any time. (This is the technique of 'diversity' well known in long-distance radio links.) But whatever we achieve technically, the final limitation is the fact that there is only one moon, and this one moon will almost certainly not be visible from both terminals of the communication link, just at the time when the link is most needed. This is a problem which is rather outside astronomy and is in reality a challenge to the users of radio communications to adjust their demands to fit in with a timetable depending on moonlight rather than sunlight. The possibility of moon-echo links is certainly there, and in the present difficult state of radio communications there can be little doubt that a use will be found. Yet again basic scientific research will pay a practical dividend of a quite unexpected kind.

The Distance of the Moon

The last part of the story of moon radar takes one's thoughts back to classical astronomy. The distance of the moon is not a yardstick by which the size of the universe is measured, as is the distance from the earth to the sun, but it is a distance which can be related directly to the size and shape of the earth. Classically the distance was found by a triangulation method (Figure 48), in

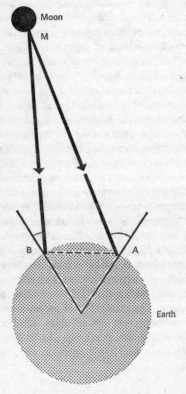

48. The distance of the moon found by triangulation. The separate distances AM and BM can now be found by radar

which the baseline was a large measured distance on earth. The determination of this distance requires the size and shape of the earth to be known; the angle between directions of the moon seen from either end of the baseline then gave the distance quite simply.

The distance to the moon now can be measured by the time taken for a radar pulse to reach the nearest part of the surface and return to the transmitter. This involves accurate time-keeping and also a powerful and sharp radar pulse, so that its arrival back can be precisely timed. The velocity of radio waves *in vacuo* is known, but some slight slowing-down of a radar pulse might occur in the ionosphere. These difficulties can all be met, and an accuracy of one part in 10 million has already been achieved, corresponding to an accuracy in distance of a few tens of metres.

Remembering now that the radius of the earth is 6 000 km, it is clear that to find the distance between the centres of the earth and moon by radar involves a knowledge of the shape, or 'figure', of the earth. A radar observation at one particular place can be used to determine the distance of the moon if the distance of that place from the axis of the earth is known; conversely it can be used to find the distance from the axis if the moon's distance is known. A possibility exists here of geophysical ex-ploration via the moon, whereby radar stations all over the world can combine results to produce a new determination of the figure of the earth. This possibility is in practice being taken over by a new form of radar, using light pulses from a powerful laser reflected from targets placed on the moon by the astronauts. The accuracy of distance determinations with this new laser radar has already been improved to less than 1 m. It seems possible that the geophysical work using moon radar with laser light will eventually include an actual measurement of continental drift, for which an accuracy of a few centimetres should suffice for a positive measurement to be made within about five years.

CHAPTER 14

The Planets

THE planets and their movements have been familiar to men for so many hundreds of years that it may seem surprising that so little is known about their composition and, in particular, about their surface conditions. Astrologers have always assumed a position of authority in planetary matters, and in recent years the fertile minds of science-fiction writers have been filling in many of the gaps in our knowledge with fascinating detail. The scarcity of hard facts gives them enormous scope.

Astrophysicists like to point out that we know less about the interior of the earth than we know about the interior of the sun, or indeed of any other star. They rightly emphasize not so much the limitations of geophysics but rather the astonishing success of recent theoretical calculations of the generation and radiation of heat out from the central furnace of a star; the comparison with our terrestrial ignorance is startling nevertheless. When we come to consider the planets the situation is much worse. It cannot be claimed that radio astronomy greatly improves this situation, but it does play a more important part than might have been expected a few years ago. Much of the pre-radio knowledge of the planets comes from the study known as dynamical astronomy, or celestial mechanics, whereby from the motions of the planets in their orbits round the sun their masses can be found, and combining mass with diameter the density can also be found. From the density and appearance, some idea is then obtained of their composition. This is complemented by spectroscopic study of the composition of their atmospheres. The detection of radio waves from some of the planets adds a small, perhaps speculative, amount to the knowledge we have of the planets, but holds promise of further discoveries as sensitivity in receiving equipment is improved.

If we compare the inert sphere of the earth with the dramatic explosions of stars and the collisions of galaxies in outer space, we could hardly expect any radiation from the earth to be significant in comparison with theirs. The planets are, however, quite warm enough to generate thermal radio waves which can be used to measure their surface temperatures, and indeed one planet produces radiations that are rather more spectacular than the dull background of mere thermal radiation. Even though the main contribution of radio observatories to planetary research will in future be the collection of data telemetered back to earth from planetary space probes, the reception of natural radio emission and the use of new radar techniques still provide most of our present knowledge of physical conditions on the planets.

Planetary Radar

It may seem strange that much of our knowledge of the interior of the planets, though not of their surfaces, comes from dynamical astronomy. Both the mass of a planet and its volume can be deduced from observations of the planet's orbit round the sun and of any moons or satellites around the planet itself, and from these a measure of its average density is made. For some, the very low density implies that most of the observed spherical object must be an extended atmosphere surrounding a much smaller rocky core, and interesting differences in the constitutions of the planets have been shown up in this way. Venus, for example, has been found to have a density approximately 5·2 times greater than that of water, which is very close to the 5·5 observed for our own earth, while Jupiter, Saturn, and Uranus and Neptune all have densities between 0·7 and 1·4 times that of water. In their case, any solid rocky material which they possess must be in a small dense core, surrounded by ice, water vapour, ammonia, carbon dioxide, and other light-weight materials.

The success of moon radar led naturally to attempts to obtain radar echoes from the planets. As for the moon, there were the twin objectives of measuring the distances to the planets and exploring the surface conditions. The measurement of distance contributes directly to celestial dynamics, since all previous work depends on the comparatively inaccurate extension of the triangulation method used for the distance of the moon (Chapter

13). There are, however, some difficulties in obtaining sufficient sensitivity, as will be seen from the Table.

	Minimum distance (km)	Diameter (km)	Required sensitivity
Moon	380 000	3 480	1
Venus	37 million	12 400	5 million
Mars	53 million	6 870	80 million
Mercury	77 million	5 000	600 million
Jupiter	590 million	140 000	3 000 million

The distances and diameters of the planets are used in this Table to calculate the improvement in performance required over that necessary for obtaining an echo from the moon, assuming that each planet reflects as well as the moon. Formidable as the lowest factor of five million may seem, echoes from Venus were obtained by the Goldstone tracking station in the USA on 10 March 1961, followed closely afterwards by Jodrell Bank, and by the Millstone Hill radar of MIT. The American radars later went on to make successful contact with Mars and with Mercury.

To achieve this *tour de force* requires the largest possible radio telescopes and the most powerful transmitters. A sensitive and complex receiver must also be used for detecting echoes; large Doppler shifts are expected, and the receiver must allow for these. Long averaging times are used; just as Bay in Hungary only obtained definite echoes after half an hour of averaging, so it may take some hours of work to build up a clear echo from Mercury.

An example of the powerful apparatus needed is the radio telescope at Arecibo, in Puerto Rico, designed specifically for radar observations. It uses a hemispherical reflector 1 000 ft across, and a transmitter with a mean power of 100 kilowatts. The field of radar astronomy is indeed an expensive one, and not one in which many competitors are found.

Systematic observations of Venus and other planets have been very rewarding. As with the moon, the surfaces of the planets can be explored by measuring their reflection coefficients and by observing the fading of the echo. The results may best be compared with conditions on the moon. About three quarters of the moon's surface reflects as though it was completely smooth, except for the very shortest radar wavelengths which show up a roughness on a scale of about 1 cm. The average reflection coefficient of the moon's surface is about 15 per cent, which is the amount expected from powdered rock. The smooth surface is tilted at angles commonly in the range of 5° to 7° from horizontal. The craters often give a stronger echo than the smooth regions, probably because they are covered with rocks rather than dust.

Conditions on the surface of Venus seem to resemble fairly closely those on the moon. Rather more of the surface is smooth, but there are some unidentified features which are very good radar reflectors. The surface slopes are a little less, about 4° to 5°. The main difference in the radar echo comes, however, at short wavelengths, where the reflection becomes much weaker; the reflection coefficient falls to only 2 per cent at 3-cm wavelength. This is not a surface effect, but an absorption in the Venusian atmosphere. The atmosphere is known to be very dense; it consists largely of carbon dioxide, which is a good absorber of short-wavelength radio waves.

The surface of Mars has very varied reflection properties, with the best reflections coming from the optically dark regions. The slopes of the surface are only about 3°. The radar reflections come mainly from a small area facing the earth, and there are variations of range observed as the planet rotates, corresponding to a variation of the surface height. Even though the surface slopes are quite gentle, there are height variations of over 10 km, close to the difference between the ocean depths and the highest mountains on earth.

Mercury is less thoroughly studied. It has no obvious individual features, but the surface is solid, and reflects rather like the moon's surface.

The surfaces of these three planets therefore seem to be similar to the moon's surface. It is not surprising therefore that the Surveyor photographs of Mars show craters, very like the lunar craters, and it is a fairly safe prediction that the faces of Venus and Mercury will eventually be found to be pock-marked in a similar way.

The rotational speed of the planets can be found from the rate of fading of the echoes. For Venus and Mercury the rotational speed was previously unknown, and the radar result has provided a surprise. It would be a reasonable expectation that the planets nearest to the sun would have a rotation locked to their orbital motion so that they would present always the same face to the sun, as the moon does to the earth. Mercury, however, is in an elliptical orbit, so that as seen from the sun it moves round its orbit at a varying speed. It tends to lock in with the sun at its closest approach, when its orbital motion is fastest; the radar result shows that the rotational period is in fact 50 days, whereas the orbital period is 80 days. It is odd to think of a 'day' lasting 7 hours, and a 'year' lasting 12 weeks. The seasons and times of day are inextricably mixed, adding to the discomforts of an unpleasantly hot planet.

The rotation of Venus is odd in two respects. First, it is retrograde: that is to say, it is rotating in the opposite direction to the rotation and orbital motion of most of the bodies in the solar system. Second, it is slow, at a rate of 243 days per rotation. The only sense that can be made of this rotation is that it tends to keep one face of Venus facing the earth over a considerable part of the orbit, as thought the planet was trying to 'lock in' on our planet. This is probably only a coincidence however, as the gravitational interaction between the two is negligibly small.

Thermal Radio Waves from the Planets

The planets all emit radio waves. Earth is unique in its profusion of man-made radio transmissions, which to a radio astronomer outside the solar system must make the planet appear as a bright radio star, a 'nova' scintillating with the most complex radio waveforms, and growing rapidly in intensity since broadcasting began. Jupiter is also unique in its intense bursts of low-frequency radio waves, which are emitted high in its atmosphere. But all planets, whether or not they have their own peculiar emissions, must radiate thermal radio waves, and the intensity of this thermal radiation can be used to measure the temperatures of the planetary surfaces.

It is easier to measure the planetary thermal radiation at the shorter radio wavelengths, where the background of galactic radio emission is not too strong, so that most observations of the planets are made at wavelengths shorter than 10 cm. The first achievement of measuring thermal waves from a planet was in May 1956, at the Naval Research Laboratory in Washington, where Mayer, McCullough and Sloanaker used a 50-ft parabolic reflector telescope working at 3-cm wavelength. They detected first the radiation from Venus, and by March 1957 they had established the surface temperatures of Venus, Mars and Jupiter. These measurements were a tremendous achievement, because the signals were so small in comparison with unwanted signals such as solar radiation, and with the noise level of the receiver. The receiver was so well constructed that it could detect a thermal noise-signal of only one tenth of a degree Centigrade.

Since these first measurements, receiver techniques have been greatly improved by the introduction of better amplifiers, such as the maser, the travelling-wave tube, and the parametric amplifier (Chapter 16). Noise levels have been improved by a hundred times, and the observations now reach to Uranus and Neptune. These planets are so far away that their discs only occupy a very small fraction of any radio telescope beam; their thermal radiation corresponds to a temperature of the order of 200 K, but it is greatly diluted by the ratio of the disc size to the size of the telescope beam, and the effective temperature of the signal to be measured is as low as 0·001 K (for Neptune).

The main results of these measurements are presented in the Table. The results are generally accurate to about 10 per cent, rather better for the nearer planets and rather worse for Uranus and Neptune. Most of the results refer to wavelengths of about 2 or 3 cm, but there are some interesting differences to be observed over a range of wavelengths, as can be seen for the several results quoted for Venus and the two results for Neptune. For comparison, the temperature of our planet is about 290 K.

There are also some temperatures available from infra-red

measurements; these generally fit the short-wavelength radio temperatures, leaving the long-wavelength ones as the unexpectedly high temperatures.

Simple physical laws can be used to predict the temperature to be expected for a solid planet in orbit round the sun. Practically all the thermal energy in the planet comes from the sun. The intensity of solar radiation, which decreases as the square of the distance from the sun, must be balanced by the thermal radiation

RADIO TEMPERATURES OF THE PLANETS

Planet	Wavelength	Temperature (K)
Mercury	11 cm	250–300
Venus	8 mm	250
	3 cm	550
	9 cm	630
	50 cm	500
Mars	2 cm	170
Jupiter	2 cm	150
Saturn	2 cm	145
Uranus	2 cm	180
Neptune	3 mm	88
	2 cm	170

from the planet, which varies as the fourth power of the temperature. The difference between these two power laws gives a temperature varying as the square root of the distance from the sun.

The calculations apply well to an inert body like our moon. If the moon were placed in orbit as far from the sun as Uranus or Neptune it would cool down almost to 60 K, which is very much less than the 180 K recorded at 2-cm wavelength for Uranus, or 170 K for Neptune. Evidently these planets are not quite like the moon. The difference is that the planets have atmospheres which profoundly affect both their temperature equilibrium and their radio radiation.

Venus provides a well-documented example of this behaviour. It shows in the Table with high temperatures at the longer radio wavelengths, and low at the shorter wavelengths and for infrared radiation. This striking difference is clear evidence for an atmosphere on Venus. The temperatures also serve as a reminder of the differences between the radio universe and the visible universe; the distinction happens to occur for Venus at the boundary between the radio and infra-red spectrum. Just what are we looking at with our radio eyes?

The conventional telescope view of Venus shows unbroken cloud covering the whole planet. The cover is so complete that nothing can be seen of any surface markings; nothing was known even of the rotation of the planet until the radar observations were made. Radio waves penetrate the cloud progressively deeper at longer wavelengths, and the temperature of about 600 K tells us that the surface is very hot while the atmosphere above it is cold.

It is very remarkable that the surface is at a temperature where lead melts, while the atmosphere is at a temperature well below the freezing point of water. This is explicable as a sort of greenhouse effect, with the atmosphere acting as the glass which keeps the interior of a greenhouse warmer than the outside. The atmosphere of Venus radiates away heat in the infra-red spectrum, cooling as it does so, but the blanketing effect of the atmosphere forces the temperature of the surface up much higher before it also can radiate sufficiently to reach equilibrium. Perhaps if we could emulate this efficiency in our greenhouses we could construct a new kind of solar furnace, whereby water could be boiled by the heat of the sun, and solar heat could then be used in electric generators.

Jovian Radiation Belts

The most exciting planet from the viewpoint of radio astronomy is Jupiter. Both at long and at short wavelengths it has a radio brightness greatly exceeding the thermal radiation from its surface, and it is quite difficult to sort out thermal and non-thermal radiation even in the centre of the wavelength range, at about 0·5 metre. At short wavelengths the radiation has the following characteristics:

(i) The source is wider than the visible disc of Jupiter.
(ii) The radiation is partly polarized.
(iii) The plane of polarization rocks back and forth according to the rotation of the planet.

The explanation is clear. We are observing on another planet a belt of trapped electrons, like the radiation belts surrounding the earth. The radiation is synchrotron radiation, and the polarization is determined by a large-scale magnetic field on Jupiter. The terrestrial belts are known as the van Allen belts, from the satellite-borne experiments conducted by van Allen in the USA. The Jovian belts must be very much more powerful than the terrestrial belts, and they are probably trapped in a more powerful magnetic field. The location of the Jovian belts, high above the surface of the planet, can be seen in the radio map of Figure 49.

The terrestrial van Allen belts do not produce a detectable radio signal, which is fortunate as we might find the cosmic radio signal obscured by them. However, a test of a nuclear bomb, called 'Starfish', set off at high altitude, unexpectedly injected a large cloud of electrons into the lower parts of the terrestrial belts, and this radiation was both powerful and persistent. At long wavelengths it dominated part of the sky for several months. Fortunately the radiation was only observable from low latitudes, and was not troublesome to the main radio observatories. It served, however, as a warning that experiments of this sort may do far more harm than good. 'Starfish', and other such large-scale experiments, have had as their most useful outcome some very serious international discussions on the control of experiments which affect our natural environment. We can only hope it will not happen again.

We turn now to the long-wavelength radiation from Jupiter, which also departs from thermal radiation but in a quite different way.

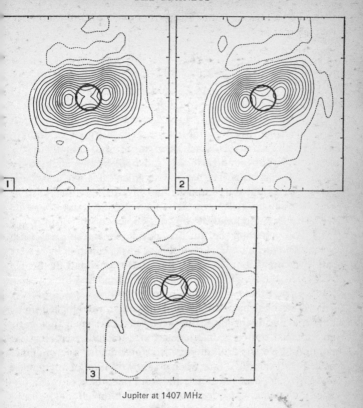

Jupiter at 1407 MHz

49. Jupiter's radiation belt. The contours of radio emission show peaks
outside the surface of the planet (shown by a circle). As the planet rotates
the magnetic field rocks and the radiation belt tilts. The three radio maps are
at 120° intervals of rotation (after N. J. B. A. Branson, Cambridge)

Radio Flashes from Jupiter

But for a fortunate accident, nothing might have been known
of Jupiter's radio flashes. They are still quite unexplained, and
were certainly not a feature of the radio sky to be expected and to
be sought out deliberately. The accidental discovery, the work of

Burke and Franklin of the Carnegie Institution in Washington, was the result of a nice piece of detective work.

These two observers were experimenting with a new type of receiving aerial of a type similar to the Mills Cross which had just been built in Australia. This aerial was designed for the rather low frequency of 20 MHz as an exploration of radio astronomy technique at the long-wavelength end of the radio spectrum. As I was responsible for the design, I may describe the aerial as being of quite crude construction, but it was specifically designed to have a beamwidth narrower than 2°, and was to be used firstly for the measurement of the intensities of radiation from several known cosmic radio sources. The aerial was difficult to align, and it was decided to leave it at a certain definite directional setting for some time, receiving signals from the same piece of sky day after day. It was midsummer, and the obvious choice for this strip of sky was that containing the sun. In midsummer the sun reached a north declination of 23°, and just near this declination there were also two bright radio nebulae, the Crab nebula and the nebula IC 443. So the aerial was set at 23° and left, and, as it happened, it was left set that way for two months.

All sorts of signals, wanted and unwanted, were picked up by the aerial during this time. But one mysterious signal, apparently an untraced source of interference, erratic, and resembling car ignition or electric drill interference, appeared for a few minutes each day. It came at a time when no known source of interference was anywhere near the aerial, and from the timing of its appearance it could not come from any fixed position in the sky. Burke and Franklin, in their detective work, noticed that Jupiter's declination at that time was quite near to 23°, and when they compared its movements with those of the unknown source of interference they found an exact agreement. To their astonishment they had the first recorded radio signals from Jupiter, without having intended to look at planets at all, and they had in fact the raw materials of a whole new branch of radio astronomy.

The announcement of this strange radiation set radio astronomers all over the world directing their aerials towards Jupiter, but with surprisingly little success. It turned out that the coincidence of the aerial setting with the direction of Jupiter was matched by another coincidence – the frequency of 20 MHz was in the centre of a quite narrow band of frequencies on which the planet was found to radiate. With an aerial of greater beamwidth the radiation would never have been distinguished from genuine interference; the combination of beamwidth, declination setting, and frequency was essential but entirely fortuitous.

It so happened that in Australia, several years previously, the frequency of 18 MHz had been used for recordings of cosmic radio waves and these records were now carefully scanned for any signs of Jupiter signals. Originally insignificant and quite unrecognized, these signals were clearly to be seen. Moreover, as the 18 MHz recordings were made over a whole year, they were searched for variations in the signals, and some of the characteristics now well established were obtained from these early records.

During a thunderstorm sharp crackles may be heard in radio receivers. The Jupiter signals seemed at first to be very like these. The records in Figure 50 show sharp spikes from pulses of about 1 second in duration. This can be observed for 1 hour or more at a time, followed by several quiet hours, just as would be expected from a thunderstorm. The energy in the storm would have to be of truly Jovian proportions for a lightning discharge to be recorded so far away; a rough calculation gives a factor of 10^{14} between the power in a Jovian flash and a terrestrial flash. This suggestion of a massive discharge must be dropped, however, as it was soon found that the signals are in fact quite unlike lightning signals.

Firstly, the spectrum is unlike any known radio spectrum. On either side of 20 MHz the power falls off to an undetectable level at about 10 MHz and 30 MHz, almost as though there were a deliberate choice of frequency, as in a broadcasting station. Secondly, the radiation is circularly polarized, just like sunspot radiation.

Since we can hardly postulate a Jupiter transmitter deliberately radiating on a 20-MHz waveband, the only possible source of these radiations would seem to be the planet's ionosphere. Here, as in our own ionosphere, each part has its own special frequency, according to the number-density of electrons, and here also the

(a)

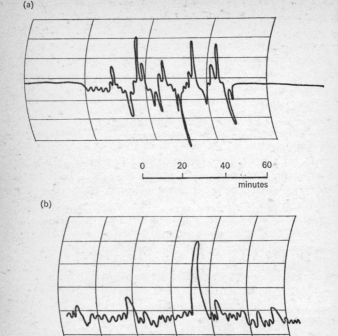

0   20   40   60
minutes

(b)

0   10   20   30
seconds

50. Radio waves from Jupiter. (a) Recording of circularly polarized signals, using an interferometer. (b) Recording of an individual pulse

magnetic field of the planet forces the electron motions into circular paths. The only questions remaining are, how does the Jovian ionosphere behave in such a way when ours does not, and also why does only one part of it do it? For as well as the radiation being limited to a rather narrow waveband, it is also limited geographically to a single source at one place on the planet.

Jupiter rotates about once every ten hours; day and night come rather more rapidly there than here. The rotation is easily observed as there are conspicuous markings, such as the famous Red Spot, which have been observed for many years (Plate XVII). Strangely enough, the periods of rotation found from watching these markings differ according to their latitude, and it seems that the visible surface acts as a liquid. Observations show that the surface near the equator rotates in 9 hours 50 minutes 30 seconds, while that at the poles takes 9 hours 55 minutes 40 seconds. The radio waves are only observed when one side of the planet faces us, and the periodicity as measured by the radio waves is 9 hours 55 minutes 30 seconds. This period places the source of the disturbance well away from the equator, and fixes its longitude to a few degrees. It must lie somewhere in a band of markings containing the great Red Spot. It would have been gratifying to find that this position coincided with the great Red Spot, but in fact it only fits with more minor markings, which may indeed have nothing at all to do with the radio waves.

Jupiter has four large satellites, or moons, which can easily be seen with the aid of a small telescope or a pair of binoculars. One of these, the satellite Io, passes through the radiation belts of the planet, and in doing so disturbs the magnetic field and the electrons trapped in it. It turns out that the long-wave Jupiter signals are profoundly disturbed by this, and a variation with the orbital motion of Io is responsible for much of the variability of the radio emission. This link between the radiation belts and the radio flashes shows that the short- and long-wavelength radio emissions must be considered as closely linked phenomena, both depending on the electrons trapped in the radiation belts, and both driven ultimately by the power in the solar wind of charged particles streaming out from the sun.

Do other planets have similar radio emissions? As far as we know the answer is 'No'. Sharp radio impulses from Venus have been recorded by one observer only, and without verification these recordings must remain in doubt. As Jupiter has such an interesting source of radio waves, we might expect the other large planets, Saturn, Uranus, and Neptune, to radiate also. Saturn is a likely looking candidate, with the similarity of its atmosphere to that of Jupiter and with those spectacular and quite magical rings round it we feel we could expect almost anything from it. But according to results so far, we must regard Jupiter as an exception, and we must hope that its radiations are no mere temporary phenomenon which will fade and be lost like other surface features. Even the great Red Spot is only eighty years old, and when so little is known about the appearance and disappearance of such large markings we may one day find ourselves lamenting the loss of the only known planetary pulse transmitter.

CHAPTER 15

Meteors and Comets

ON any starlit night the unchanging majesty of the heavens may be suddenly enhanced by the flashing trail of a meteor or shooting star. A patient observer may be rewarded by the sight of perhaps half a dozen in an hour; during a few nights of the year he may see over a hundred in an hour, and when this occurs we speak of the increased incidence of meteors as a 'meteor shower'. Meteors may be seen from all over the earth and are continually shooting into our atmosphere. In a single period of twenty-four hours there must be over a million meteors falling on the earth as a whole and over 20 million during meteor showers.

There are also many more, fainter, meteors, visible only with the aid of a telescope where their trails are often found crossing photographs of stars. Some of the largest meteors survive their passage through the air and fall to the earth as meteorites, the largest of all weighing some hundreds of tons. Some of the very smallest also fall to earth without burning up completely and can be found as a light meteoritic dust. In all, about 10 tons of material are added to our planet every day, by the impact of meteors.

Fragments and dust from meteors and meteorites show that they are made from materials similar to those of which the earth is made, although there appear to be two distinct types of meteor, the stone and the iron varieties. The stone ones usually contain some iron, but on the whole their composition corresponds closely with the outer rocky parts of the earth; the iron meteorites contain up to 99 per cent of a nickel-iron alloy. Probably these correspond with the inner parts of the earth, which are known to be heavier, and to be magnetic. Here is our most tangible connection with outer space; so close in fact that it was for a time considered that meteorites actually came from earth, being shot out from volcanoes in prehistoric times.

Meteorites may be seen in many museums of natural history. They range in size up to the 36-ton specimen from Cape York,

Greenland, now preserved in New York, and all of them have ended on the surface of our planet after a journey that aspiring space travellers might envy but could not hope to imitate. Where do they come from?

Considering that the origin of the solar system itself is an unsolved mystery, that of the meteors cannot be expected to be obvious. The question that must first be answered – and here radio astronomers have made an important contribution to meteor astronomy – is whether or not they belong to the solar system themselves. If they do not, do they permeate interstellar space throughout the galaxy? And if they do belong to the solar system, do they show the concentration to the plane of the ecliptic which is the most marked feature of the other members, planets, asteroids, and comets?

Comets and Showers

In searching for the origin of meteors we distinguish at the outset between the meteors arriving in showers and those of sporadic occurrence. Sporadic meteors have long been a puzzle, but the origin of the meteor showers has become quite clear since the discovery that they are closely associated with comets (Figure 51).

Comets which can only be observed in telescopes are quite common. The Comet Arend–Roland which appeared in 1955 was much more spectacular, and provided a world-wide spectacle equalling Halley's Comet, which last appeared in 1910. Comets such as these move rapidly across the sky, coming into sight as a faint diffuse point and developing the characteristic tail as they approach the sun. As they recede, after only a few days, the tail swings round so that it always points away from the sun, just as though the sun were continually blowing it away from itself.

The motion of a comet can be plotted with great accuracy, and an orbit can be computed just as for a planet. Most comets prove to lie in elliptical orbits, although the ellipse is often so elongated that the portion near the sun is very like a parabola. On a hyperbolic orbit, a comet would leave the solar system altogether, and it would then be a passing visitor, seen once only. A parabola is a 'border-line' orbit, with elliptical and hyperbolic orbits as general cases on either side. Cometary orbits go so far out from the sun that the period of return visits may extend to hundreds of years, and perhaps more, but many periods shorter than this are known.

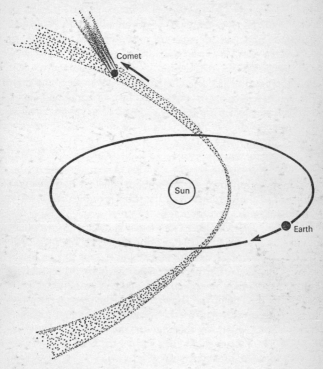

51. The origins of meteor showers. The earth in its orbit encounters a trail of debris following the orbit of a comet

For the recurring comets, measurements of the orbit on one single passage often give a value for the period which is closely corroberated by successive visits of the comet. Halley's Comet is the prime example of this, with twenty-seven appearances at regular intervals of seventy-seven years since 87 B.C. There are also several comets with a period of less than ten years, moving in much smaller orbits. These are believed to have been deflected from larger orbits by passing close to the massive planet Jupiter.

A passage close to the sun can have a disastrous effect on a comet. A comet seen in 1845 was one that had been seen on several occasions; in 1826 Biela had computed its orbit and shown that it had a period of $6\frac{1}{2}$ years. On this occasion, however, it was seen to be accompanied by a small fellow-traveller, which appeared to grow in brightness during the passage of the comet, as if the head were breaking into two parts. The next return of Biela's comet, in 1852, was the last time it was seen, and it then consisted of two faint parts separated by over a million miles. On several subsequent periods it was looked for, and no comet was to be found at all, but the remains of the comet were quite clearly seen. At the appointed times of return, in 1872, 1885, 1892, and 1899, when the earth crossed the orbit of the vanished comet, there were spectacular meteor showers. Apparently the debris of the comet was still travelling round the same orbit, and when the earth passed through this orbit the shower of dust was observed as the Bielid meteor shower.

Giacobini's comet also has a meteor shower, the Giacobinids, and in this case the shower does not merely represent a sort of afterlife of the comet, as both comet and meteor shower exist together. Our planet, Earth, crosses the Giacobini orbit at different distances from the comet each year, and it appears that when it is nearest to the comet the shower is most intense. Certainly the largest meteor showers of this century came during passages closest to the comet. Another shower, the Perseids, has the same orbit as that of a comet, and this shower has been recurring regularly for at least 1 000 years.

Meteor showers are therefore seen to be the results of collisions between the earth and the orbits of comets, orbits which contain either the stuff from which comets are made or the debris into which they disintegrate. Comets and meteors then constitute one problem together, which is a part of the general problem of the origin of the solar system.

Sporadic Meteors

It is one thing to determine the orbits of a group of some hundreds of meteors, all of which prove to be travelling along the same elliptical path, and quite another to determine the orbits of the few sporadic meteors seen every night. There is now no relation between successive meteors; they all travel in different directions. Computing orbits of individual meteors, appearing without warning in any part of the sky, would seem to be an almost impossible task, but it is one which has fascinated many astronomers, and much work has been done in this field. Professor Whipple in Harvard has done extensive work in photographing meteor tracks in such a way as to determine both the speed and direction of the meteor. In 1945 when Hey and Stewart first used radar methods for studying meteor trails, the difficulties in this study by photographic methods were still so great that no definite conclusions could be drawn. Some observers thought that no hyperbolic orbits were to be found, others that all sporadic meteors had hyperbolic orbits and came, not from our own solar system, but from the remote spaces between the stars. Radio astronomy, or rather 'radar astronomy', has now firmly settled this controversy.

Radar Echoes and Meteor Trails

When a meteor strikes the atmosphere it may have any velocity between about 8 and 80 km sec^{-1}. Encounters between atmospheric gas and a meteor result in a slowing down of the meteor and also in the heating which makes it glow visibly. The trail of such encounters is a trail of ionized gas, which diffuses away and recombines over a period of some seconds. This trail reflects radio waves. The details of this reflection have been investigated for a number of years, and two types have been found. Around the meteor itself is a head of dense ionization, which can produce its own echo. The long trail acts separately from the head, and as far as long radio waves are concerned it looks like a long thin piece of conducting wire stretched across the sky. Such a piece of wire reflects radio waves back upon their own path when the waves are reflected perpendicularly to the wire, at 'normal incidence'.

This second type of reflection was the first to be demonstrated and understood. In 1945 Hey and Stewart, working in the Army Operational Research Group, arranged three radar sets using the same long wavelength all to search simultaneously for meteors in one patch of sky. The radars were situated many miles apart, at Richmond, at Walmer, and at Aldeburgh, so that this small patch of sky, at a height of about 96 km, was being examined from three different directions. Each radar received over 2 000 meteor echoes during the six weeks of observations, but only on four occasions was the same trail observed by two sets simultaneously. Each radar was selecting only those meteor trails which crossed the beam at right angles.

During a meteor shower the same three radars showed an echo rate which reached a peak at different times, when the shower was cutting perpendicularly through the separate aerial beams; these times gave a measure of the direction of the shower, or its 'radiant'. It was now possible to measure the radiant of a shower without a single meteor being seen, and further, it was now possible to measure the radiants of meteor showers falling on the earth in daylight hours. Hey and Stewart in fact showed in the same year that there were important showers, to be observed only by radio methods, through many summer days, although it was the work of Lovell and his colleagues at Jodrell Bank which gave the first clear description of these daytime showers.

In October 1946 several different teams were already at work on meteor echoes. At Jodrell Bank the new radio observatory was making a modest beginning with one mobile radar set in a trailer. Appleton and Naismith at the DSIR research station at Slough, and new research teams in America, were all preparing to join with Hey in observations of the meteor showers expected when the earth crossed the orbit of the Giacobini comet on 9–10 October. A prolific meteor shower was expected at this time, but the number of echoes observed was far greater than ever was hoped. The sporadic rate of echoes on one particular equipment was about two per hour: this was observed up to the time of the expected shower when the rate jumped up to 200 per minute, rising to this very high rate and falling again in only six hours. From that time onwards radar astronomy was an established science, with all kinds of new techniques rapidly being added to the straightforward military radar sets. New results poured out at a pace which seemed to be limited only by the speed of the earth's passage through successive meteor showers.

Meteor Velocities

The elucidation of the origin of the sporadic meteors depended on the measurement of their velocity. There are now several ways of measuring this velocity: the most obvious apply to the type of echo which comes from the head of ionization rather than from the trail itself. This head can be observed as a moving echo, like the radar echo from an aircraft, and the velocity can be measured either from the rate at which its range changes with time, or, more directly, from a measurement of Doppler shift in the echo. Strangely enough, the echoes from the line trail can also give a measurement of velocity, from a peculiar variation of the echo as the trail grows in length. These variations are shown in Plate XVIII. Diffraction theory must be used to explain them; fortunately the precise theory was already available, having been formulated by Fresnel for light waves in 1816. According to Fresnel's diffraction theory, the echo builds steadily up to half its final value at the time when the meteor crosses the foot of the perpendicular from the radar set to the meteor trail. When the trail has crossed this point, oscillations of echo amplitude occur as the path length to the meteor increases by successive increments of one wavelength.

In 1947, 1948, and 1949, successful velocity determinations were made for the Geminid meteor showers by this diffraction technique, at Jodrell Bank. A velocity of 36 km sec^{-1} had already been found for several meteors of this shower by Whipple, and the radar measurements gave precisely the same result. Now the radar measurements could be used to tackle the unsolved and burning question of the sporadic meteors.

When a meteor is encountered by the earth travelling in its orbit round the sun, the actual velocity with which the meteor enters the earth's atmosphere depends on the relation between the meteor orbit and the earth's orbit. If the meteor is met head-on, for example, the velocity may be three times that of a meteor which is overtaking the earth in its orbit. For each direction there is, however, a critical velocity which represents a border-line between theories, with higher velocities belonging only to meteors which must have travelled in hyperbolic orbits, entering the solar system from outside.

The idea of encountering meteors from outside our solar system is one that excites the imagination, and many experiments have now been performed to search for such high-velocity visitors. It is perhaps rather disappointing that few have been found which can even be suspected of such a wild journey. Figure 52 shows the result of one search, when the meteors arriving 'head-on' to the earth's motion were being investigated by J. G. Davies and A. C. B. Lovell at Jodrell Bank. The smooth curve shows the distribution of velocities to be expected if the sporadic meteors were all travelling at the limiting velocity of a parabolic orbit, and it is evident that the actual velocities are well below this limit. Sporadic meteors arriving from many directions have also been investigated by D. W. R. McKinley in Canada, and all observatories agree that practically all these meteors belong to our solar system. Further, it is becoming clear that many of these meteors move in orbits rather similar to those of the planets, and it begins to seem rather that sporadic meteors are closely related to the asteroids, the group of tiny planets, some no more than a few miles across, which occupy the space between Mars and Jupiter. It may be that the smallest asteroids are just the same bodies as the largest meteorites, and that all are in fact simply planets, with a continuous gradation in size. But to establish the exact relation between the various small members of the solar system, comets, meteors, and asteroids, we need to know more about the distribution of the sporadic meteors in space. This is now the main subject of meteor research. The most puzzling feature of the distribution is the appearance of some meteor orbits inclined at large angles to the plane of the ecliptic; the planets and asteroids are so closely confined to the ecliptic that it is hard to find any dynamic reason for this departure.

With meteors, asteroids, and planets forming such a diverse collection of objects travelling in the sun's gravitational field, it is interesting to speculate whether the smaller bodies are planets in the making or planets in a stage of dissolution. In fact which came first, the planets or the meteors, or does the process even work both ways? There is so much detailed evidence that must be taken into account by any theoretical explanation of the

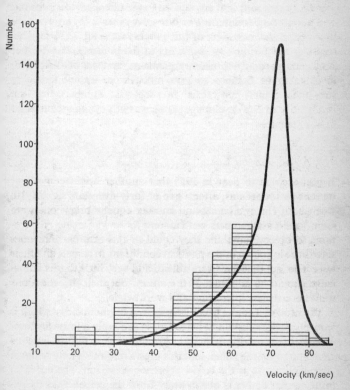

52. The measured velocities of meteors. The curve gives the expected distribution of velocities if all meteors have the energy required to escape from the solar system. The 'histogram' of observed velocities shows that meteors all have energies lower than this limit (after J. G. Davies and A. C. B. Lovell)

planetary system that any theory becomes liable to severe criticism. The only theories which approach a convincing explanation propound that the planets condensed out of the same nebular material as the sun, and they suggest that the formation of planets of all sizes would be a natural occurrence in this process. This would mean that the meteors and asteroids represent condensations which never achieved sufficient size to grow to a true planet.

It cannot be said that any theory so far offers precise information about the development of our solar system. When we come to a better understanding of this process we shall be better able to assess the position of humanity in the universe. For if other stars tend naturally to acquire planetary systems of their own, invisible to us because so very much more distant than our sun, then planets must exist in such vast numbers that the probability of life developing in the warmth of other suns is no longer remote.

The Arend–Roland Comet

It was exciting to hear in 1955 that another bright comet was on its way towards us, after a gap of forty-five years since 1910, when both Halley's comet and another equally bright one were seen. Radio astronomers were among those who made preparations to observe anything they could of this rare phenomenon, since little is known of the physical conditions in comets and their tails. The fact that they were able to add very little to this knowledge should be admitted at the outset, but their investigations were nevertheless interesting and worthwhile.

The head of a comet usually contains a bright nucleus, which is the only really substantial part. This is probably a loose agglomeration of meteoritic material. Around it is a diffuse gaseous envelope called the coma, and this draws out into a long tenuous tail which grows as the comet approaches the sun. The tail is so insubstantial that it is blown away from the comet head by the mere pressure of the solar wind, so that the comet always appears to be pointing towards the sun. The light from the comet is reflected sunlight. Faint spectral lines in it indicate that it contains molecular material with low atomic weights.

The radio astronomers who were joining in the observations of Comet Arend–Roland (named after the Belgian astronomers who first discovered it) had to ask themselves what radio studies could best contribute to our understanding of the comet. Their studies would be limited to about two weeks of observations and careful preparation was necessary to make the best use of such a limited time for experiment and observation.

Certainly the head would be no more accessible by radio astronomy than are the asteroids or the more distant planets, but the tail might well contain ionized gas in sufficient quantity for appreciable refraction of radio waves, and it might even radiate at radio wavelengths if there were enough electrons at a high enough temperature. Two kinds of experiment were planned therefore: one to detect direct radiation from any part of the comet, and the second to examine radio waves from more distant sources when they traversed the comet's trail.

Radio telescopes of all kinds were trained on the comet during the week of its closest approach. There were several reports of radio waves emitted from the tail, but there were very many more which stated categorically that nothing detectable was emitted over a wide range of frequencies. Only one positive report remains, claiming that radiation was received on 600 MHz, using a 27-ft parabolic reflector. In the face of the many negative reports this had to be considered as doubtful, and the explanation that the 600-MHz emission was a radio spectral-line emission cannot easily be accepted. The search for direct radiation was therefore inconclusive; but subsequent comets, such as Comet Bennett (Plate XVI), confirm the result that comets are not radio emitters.

As for the refraction measurements, these proved also to be rather disappointing. Few radio telescopes were available to work at the low frequencies required, and only one experiment was performed. G. R. Whitfield at Cambridge, with his 38 MHz radio-star interferometer was unable to show any evidence of refraction, so that the sum total of radio observations of the comet gave it a very poor radio performance. Whitfield's results could be interpreted as showing that there must be less than 10^4 electrons cm^{-3} in the comet's tail – but it would have been surprising to find that amount anyway.

It might appear that radio astronomers were wasting their time in looking at so unpromising an object. But when the question is asked 'Do comets radiate directly, or can they refract radio waves?', the answer 'No' is just as much an answer as a more positive one, and an inconclusive result makes us seek further evidence when another opportunity arises. It can hardly be stated as scientific fact, but sometimes it would seem that the radio significance of celestial objects is inversely proportional to their optical impression – and the Arend–Roland comet certainly was no disappointment optically.

CHAPTER 16

Radio Telescopes

THE survey of the radio universe has now been followed in the preceding chapters from the sun out to the edge of space and time, and back through the moon and the planets to the solar system. We now return to the instruments with which this exploration has been carried out. What are radio telescopes, the 'radio eyes' which have so greatly extended our view of the universe?

The range of wavelengths of light which can be appreciated by the human eye lies in the centre of the range of wavelengths which penetrate the terrestrial atmosphere. On either side of this visible spectrum lie the ultra-violet and infra-red regions, which are not greatly absorbed by the atmosphere. In these regions the use of photographic methods allows a considerable extension of astronomical observation, by techniques which closely correspond to normal vision. It is not impossible to imagine the actual extension of normal vision into the ultra-violet and infra-red ranges by some biological development, though this would be a much more complicated matter than, for example, the extension of the audible spectrum of sound to include the high frequencies of sound waves used by many animals smaller than ourselves.

The development of human eyes for the reception of radio waves, which penetrate the atmosphere equally well, is, however, quite inconceivable. It is not impossible to find suitable refracting materials in nature: paraffin wax, for example, can be used for making lenses for radio waves though it could hardly be considered an ideal material for eyes. The construction of a retina or other sensitive element is also hard to imagine. But the real difficulty lies in the much longer wavelengths of the radio spectrum, which make the accommodation of a directive aerial array, or antenna, a difficult problem. In fact, in any reasonably sized animal the aerial system would probably determine the appearance of the animal to the exclusion of all other features.

The evolution of eyes sensitive to light is marvellous indeed, but our blindness to radio waves is entirely reasonable. We have then to rely on our man-made aids to vision, the radio telescopes, if we are to take advantage of the radio window to the universe.

The history of astronomy begins with a long introductory era of observations unaided by any sort of telescope. In those early years astronomical instruments were all devices concerned entirely with the measurements of the positions of heavenly bodies, and of time. All over the world the remains of observatories belonging to ancient civilizations remind us of the importance of systematic observations, directed usually towards the magic of astrology or the necessities of navigation. The classical peak to which pre-telescopic astronomy attained is found in the work of Tycho Brahe, the Danish astronomer of the seventeenth century. His records of the movements of the planets allowed Kepler to formulate general laws governing the movements of planets around the sun. Both Tycho Brahe and Kepler also had the distinction of observing supernova explosions, and the remains of both these are now radio sources (see Chapter 6).

The use of telescopes enabled this positional type of observation to be made with much greater accuracy, and at the same time it opened the way for a less mechanistic and more astrophysical approach in astronomy.

Radio astronomy has no 'naked eye' stage of observations. From its earliest beginnings it has depended on radio telescopes, and the approach, all along, has been entirely physical. The development of radio astronomy, and that of its necessary instruments the radio telescopes, has not been determined by practical requirements of navigation or the dictates of those who determine 'what the stars foretell'. It is true that the new science can be of some aid in the new navigational problems of space travel; but a former Astronomer Royal put this into perspective with his dictum, 'Space travel is bilge.' And as for any concern with astrology, we have yet to see a horoscope cast with the aid of the radio stars.

Positional work has nevertheless been of great concern in the early work in radio astronomy, and, in this field, despite the difficulty of the longer wavelengths to be contended with, we find ourselves well ahead of Tycho Brahe in the accuracy we are able to attain. Accuracy is now sufficient for most problems of identifying radio stars; but this attainment has come only after the development of a series of radio telescopes specifically built for positional work.

Side by side with position-finding in radio astronomy has gone the development of radio telescopes of great sensitivity. Here the criterion is 'the bigger the better'. The first radio telescope, made by Reber, was designed solely for great sensitivity, just to see what could be seen in the sky. In Reber's maps can be seen the first delineation of the radio waves from the Milky Way, and also some unresolved patches which later were shown to be 'radio stars'. From Reber's radio telescope derives a line of radio telescopes with increasing sensitivity, some designed specifically to detect as many radio sources as possible, and others for more general purposes.

It is impossible to separate radio telescopes into strict categories of sensitive and position-finding instruments, but there is a broad distinction between the two types. The distinction has led to quite different techniques of construction, each chosen for a specific experimental purpose. This deliberate choice of a particular construction for a particular experiment has meant that many of the notable radio telescopes of the last two decades are already obsolete, with their experimental purpose fulfilled; no single radio telescope can ever claim to be a truly general-purpose instrument. Some of the large new radio telescopes have much greater versatility than earlier constructions, but it is recognized that they are still limited in their range of use. Versatility implies simplicity, and the parabolic reflectors, which are the simplest and the nearest approach to general purpose instruments, will therefore be described first. Interferometers of increasing complexity follow, leaving the actual radio receivers until last. This order is analogous to dividing the subject of optical telescopes into sections on large reflectors, on spectroscopes, and on photographic plates.

Parabolic Reflectors

The parabolic reflector of an electric fire receives heat from the element at its focus and reflects it forward over a wide area. The

221

problem of detecting radio waves from a point in the sky is the opposite problem of collecting the radiation falling on as large an area as possible, and concentrating it at one place so that it can be fed into a radio receiver, amplified, and recorded. A parabolic reflector with a receiving aerial at its focus lends itself perfectly to this problem.

The 200-in diameter optical telescope on Mt Palomar is built round a parabolic mirror which collects light falling on an area of 20 m² and concentrates it on a photographic plate at the focus. This 200-in mirror could also be used for collecting radio waves, but it would be a rather wasteful use of so accurate a surface. This surface is true to a fraction of a wavelength of light, and this is about a million times as good as it need be for radio waves. The parabolic 'mirror' used for collecting and focusing radio waves can be much larger than the 200 in of the optical instrument, according to the wavelength of the radiation it is to receive, and can be made of sheet metal or a mesh of metal and wire, accurate to about a tenth of the wavelength concerned.

Apart from the obvious differences in construction and precision of surface between the optical and the radio telescope, there is an important difference in the arrangements at the focus. An optical telescope forms images of all stars within a considerable field of view, whereas a radio telescope has only a single pick-up for radio waves, at the exact focus of the parabola. The information about the sky comes not as a picture or a map, but as a voltage; furthermore this voltage represents not the radiation from one single direction but rather the average over a whole range of directions determined by the resolving power of the telescope.

Plates XX and XXI show parabolic-reflector radio telescopes of various sizes. So many of these 'dishes' are now in use throughout the world that it is impossible to show more than a representative selection; however, it is not only the number but the size which is increasing. In fact it would seem that however large a parabolic aerial a radio astronomer has available, he always has good reasons for wishing it even larger.

The 'resolving power' of a telescope is the measure of its ability to distinguish two sources of radiation separated by a small angular distance, and may be loosely defined as the minimum angular distance at which this is possible. It is also to be thought of as the angular beamwidth of the telescope, and its value is determined by the ratio of the telescope aperture to the wavelength. For example, the 200-in telescope on Mt Palomar has an optical resolving power of about 0·1 second of arc; the 250-ft Jodrell Bank radio telescope has a resolving power of 1 degree for a radio wavelength of one metre. The need for very large apertures in radio telescopes hardly needs emphasizing. Resolving power for the astronomer determines the fineness of the brush which he can use to paint his picture of the sky.

The construction of large parabolic radio telescopes presents considerable engineering problems. The reflecting surface of some large ones has been made immobile, a very economical expedient but one which limits the usefulness of the telescope. We will consider the fixed and movable reflectors separately, although some of the design problems are common to both.

Steerable Parabolic Reflectors

The engineering problem here is simply stated. A large surface, accurately constructed out of sheet metal or wire mesh, has to be maintained in its correct shape while it is directed at any desired part of the sky. The distortions which must occur as it moves, and even when a moderate wind blows on its surface, must be kept down to less than one tenth of the shortest wavelength for which the reflector is used. However stiff the structure that holds the surface in place, it will bend under its own weight, and design has to stop at the point where extra stiffening adds further weight, which itself adds further distortion.

The Table lists some of the outstanding parabolic reflectors so far built, giving their diameters and the smallest wavelength for which they can be used with full resolving power. At shorter wavelengths they are still usable although with reduced efficiency. Part of the versatility of the large parabolic reflector as compared with other large radio telescopes lies in the ease with which the wavelength of operation can be altered. They are, however, more likely to be used at the shortest possible wavelength which is determined by the precision of the structure.

Telescope and location	Diameter (metres)	Shortest wavelength for full efficiency	Beamwidth at shortest wavelength
250-ft Mark IA, Jodrell Bank	76	10 cm	5′ arc
210-ft, Parkes, Sydney, Australia	64	5 cm	3′ arc
100-m, Max Planck Institute, Bonn	100	5 cm (centre part 2 cm)	2′ arc
140-ft, National Radio Astronomy Observatory, Green Bank, USA	43	2 cm	2′ arc
36-ft, National Radio Astronomy Observatory, Kitt Peak, Arizona	11	1·5 mm	1′ arc

The performance of these instruments represents an improvement over a period of fifteen years in two different ways. The angular resolution, or beamwidth, has been progressively reduced, while at the same time the shortest wavelength at which the telescope can be used has been reduced from about 10 cm down into the millimetre wavelengths. The penalty for the extension to the shortest wavelengths has been a reduction in size.

Paraboloid reflector telescopes are now seen to be dividing into the smaller, millimetre-wavelength instruments, represented by the very successful pioneering 36-ft telescope at Kitt Peak, and the larger centimetre-wavelength instruments such as those at Jodrell Bank, Parkes, and Bonn. The problems of building accurate reflector surfaces are divided in the same way: for smaller surfaces the surface distortions are limited mainly by the effects of wind pressure and temperature gradients, while for the larger surfaces gravitational deformation plays a more important part.

Most radio telescopes are mounted on an altitude-azimuth mount, and not on the polar mount familiar in optical astronomy (Figure 53). The main advantage is that the gravitational deformations act in one direction only, since the surface does not rotate about its axis. There will nevertheless be deflections of several centimetres in the structure of a large radio telescope, and the performance will suffer unless these can be compensated for in some way. One way is to design the structure so that when it is deformed by gravitational forces it takes up the shape of another paraboloid, with a different focal point. Then a movement of the receiver will bring the whole surface back into a correct geometry. This approach has been successfully employed in the 100-m radio telescope at Bonn, where the surface remains a true paraboloid within an accuracy of 1 or 2 mm even though deflections of over 5 cm occur in the outer parts.

Another approach was adopted in the design of the proposed 375-ft (114·3 m) radio telescope known as Mark VA to be built in Wales, which was intended to be the successor of the Jodrell Bank Mark I. In this telescope a surface accurate to 2 mm would be maintained by adjusting over 200 jack points connecting the surface to the main telescope structure. This system would be capable of great accuracy at any elevation, by making a series of adjustments to the jacks according to a controlling computer program.

The idea of an adjustable surface of this kind is already familiar to optical astronomers, who use a similar system for supporting the large mirrors of reflecting telescopes. There is, however, an unfortunate precedent in a radio telescope which was intended to be adjustable in this way, but which was abandoned during construction. This was the 600-ft telescope for the Naval Research Laboratory in the USA. The foundations for the telescope still

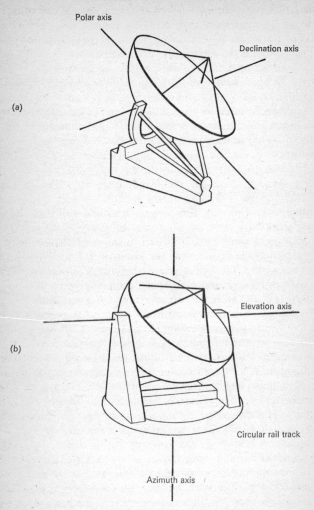

Polar axis

Declination axis

(a)

Elevation axis

(b)

Circular rail track

Azimuth axis

53. Radio-telescope mounts. Some small radio telescopes use the polar mount (a), which is usual for optical telescopes, but most use the altitude-azimuth mount (b), which can carry larger and heavier structures with less distortion

226

exist at Sugar Grove, in West Virginia, but complications in design led to its abandonment before much of the steelwork had been assembled. The total expenditure on the project had nevertheless already reached nearly $100 million.

Whatever is ultimately achieved in the construction of large steerable paraboloids, the step forward which the 250-ft reflector at Jodrell Bank represents must remain as one of the most significant in radio astronomy. The largest steerable reflector in existence at the time of its design was 50 ft in diameter; even the problems to which the new instrument is being applied were hardly defined at that time.

Fixed Parabolic Reflectors

There was for several years a second large parabolic reflector also at Jodrell Bank, built in its original form in 1947 and measuring 220 ft in diameter. Lying on its back, it looked directly up at the zenith; its beam could only be moved a few degrees when the dipole aerial at the focus was moved sideways. A paraboloid lying on its back is very much cheaper than a paraboloid mounted on towers, which can be pointed anywhere in the sky. There may well be more of these fixed paraboloids built in sizes which cannot be attained by completely steerable instruments. There is already another 250-ft bowl, used by the Naval Research Laboratory in Washington for moon radar; this is an example of simplicity of construction, as it is made by bulldozing a carefully shaped hole out of the ground, and paving it with asphalt road surfacing. The shape is right to an accuracy of 3 in, which is nearly as good as the Jodrell Bank steerable telescope, and though limited in use the great reduction in cost is commensurate with this limitation.

The size record for excavated reflectors is held by the 1 000-ft reflector at Arecibo, Puerto Rico. This is built in a natural hollow, with three towers at the rim supporting a spidery feed structure on wire ropes. The reflecting surface is part of a sphere rather than a paraboloid, and the resulting distortion at the focus is compensated for by replacing the usual receiver dipole by a long antenna system.

These fixed paraboloids cannot be classed among the most

versatile radio telescopes, but they are very well adapted to some particular experimental tasks. If directed at one part of the sky for a particular experiment it does not matter that a large part of the sky can never be seen by it. For example, in the Crimea the Lebedev Institute built a special reflector for studying the radio waves from the Crab nebula. This was a fixed parabolic reflector pointing at the right distance from the zenith, 100 ft in diameter and so accurately constructed that it worked well at a wavelength of 10 cm.

The notable 400-ft cylindrical paraboloid at the University of Illinois is more versatile, but the beam can still only be steered in one plane.

The engineering difficulties in these very large fixed paraboloids are not negligible, even if a suitable site can be found to avoid too much excavation. The main problem is access to the focus, on top of a tall tower or on a spider's web of wire ropes. Designs of fixed reflectors in which the focus is on the ground are much more attractive, but for this the paraboloid must be turned on its side, consisting then only of the part actually above the surface of the earth. The beam, which is horizontal, can be directed upwards either by tilting the reflector, or by using a second large reflecting surface which can tilt but which is flat and easier to handle than the main reflector. Examples of these were first constructed at Pulkova and at Ohio State University, and a very large tilting-plate reflector is now in use at Nançay.

INTERFEROMETERS

Let us suppose that radio telescopes can be bought by the square yard, for a price which is not affected by their shape; and let us suppose that a radio astronomer has bought a certain number of square yards of collecting surface. What shape should he construct in order to obtain the best value for his expenditure?

One fairly obvious answer is to put it all inside a circular aperture, like the parabolic reflector; this gives a 'pencil-beam' aerial, which receives radio waves from a comparatively small area of sky. But for some experiments it is advantageous to divide the total area up into two or more small parts and separate them from one another. The radio telescope then becomes an interferometer, which has a greater angular accuracy for position finding and may be used for distinguishing small objects against an intense background. A given area of collecting surface connotes a definite sensitivity; all that is changed by changing the shape is the reaction of the telescope to the angular structure of the radio sky.

An optical interferometer was made in 1923 by Michelson, who adapted the Mt Wilson 100-in telescope for measuring the diameter of stars by dividing the telescope aperture with a system of mirrors. Correspondingly the radio interferometers using two aerials are often called Michelson interferometers, and their first use was to measure the diameters of cosmic radio sources, and of the radio sun.

The first Michelson radio interferometer was made by M. Ryle and D. D. Vonberg in Cambridge in 1946, and it was used to measure the diameter of an emitting sunspot. This interferometer used two aerials, each consisting of five dipoles mounted over a sheet of wire netting. Much larger aerials are now used for the weaker signals from radio stars, but the principle of the interferometer is the same in both applications.

Michelson had already pointed out the possibility of achieving more than a simple measurement of the diameter of a star, and had shown that the whole distribution of brightness over its surface was in principle measurable by an interferometer, provided that the separation between the two components could be varied. The limit of resolving power is set only by the maximum separation of the two mirrors, or aerials, and not by their actual size. By his arrangement of two small mirrors Michelson was able to attain with the 100-in telescope a resolving power considerably better than that of the 200-in telescope. But angular resolution is not the only advantage of interferometers. In recording weak radio sources, one difficulty of a single pencil-beam aerial is the strength of the background of galactic radio emission. An interferometer can be used to reject this background and observe only sources with a small angular size, giving a greatly increased contrast in the recordings. Figure 54 shows some recordings made with an early radio interferometer where this principle was first used; it was possible on this instrument to record the radio sources with or without the background according to the arrangement of the receiving system.

Michelson's classical experiment on the measurement of diameters has been repeated many times for the radio sun and for

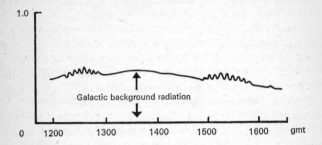

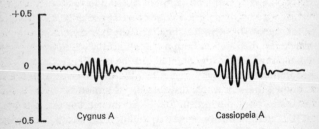

54. Early interferometer recordings of two radio sources. The top record was made with a 'total power' recording, and shows radiation from the Milky Way; the lower recording was made with a receiver which records only the discrete sources

other radio sources. Further, it has been possible to carry out his suggestion of measuring brightness distributions including measurements of the size of the radio sun at long wavelengths. No steerable paraboloid can achieve this resolution, as it would have to be over a thousand feet in diameter.

Variety in radio interferometers runs the whole gamut of variety possible in the single aerials placed at either end. Paraboloids, cylindrical paraboloids, arrays of dipoles, all are used according to the aperture and the wavelength of reception. One particular type of interferometer, however, uses only one aerial; by placing this on top of a high cliff overlooking the sea the interferometer action is simulated by the reflection of radio waves in the surface of the sea. The first interferometer radio telescope was made in this way, by J. L. Pawsey of Sydney, Australia, who placed an aerial array for 1·5 m wavelength on top of the 300-ft cliff near the entrance of Sydney Harbour and observed the radio sun as it rose over the horizon at dawn.

Again, interferometers need not be limited to only two aerials each. There are now several arrays which employ two lines of aerials, each of which has thirty-two separate paraboloid reflector aerials spaced well apart but connected to the same receiver. In optical terms this array is analogous more to a diffraction grating than to an interferometer.

Long-Baseline Interferometers

The angular resolution of an interferometer is given, in angular measure, by the ratio of the wavelength to the length of the baseline between the two aerial systems. (In units convenient for long baselines, the resolution in seconds of arc is obtained by dividing the wavelength in centimetres by the baseline in miles.) The interferometers described in the previous section mostly have angular resolutions measured in minutes of arc rather than seconds; these were found to be quite incapable of measuring the angular diameters of most radio sources, and an improvement had to be made either by decreasing the wavelength or increasing the baseline. The strength of the signals from most radio sources decreases with wavelength, and the sensitivity of receivers also decreases, so the only course open in the 1950s was to increase the baseline.

The key technical problem in interferometers is to bring the two radio signals together without altering their main character-istics of amplitude and phase. (In technical terms, this means that their coherence must be preserved, and that no separate detectors may be used.) For short baselines this may be done by connecting the telescopes with a coaxial cable, using amplifiers at each telescope to overcome the loss in the cable. The big advance on this technique, pioneered at Jodrell Bank (see Chapter 11), was to use a microwave radio link to transmit the signal from one telescope to the other, over a distance which grew by stages from 25 km to 140 km. The signal cannot be directly transmitted down the radio link without change, but must be converted to a lower frequency by mixing with a local oscillator. Preservation of the relative phase of the two signals means, however, that identical oscillators must be provided at both telescopes. This 'local oscillator' was generated at the home station and transmitted to the other end of the interferometer by another radio link.

The resolution of the angular diameters of the quasars was still not accomplished when the baseline of these radio-link interferometers grew to 140 km, and the wavelength had shrunk through stages of 70 cm, 21 cm, 11 cm, to 6 cm. The angular diameter of some quasars appeared to be less than 0·05 arc seconds, but the extension of the interferometer baselines was no longer possible. At this stage the initiative passed to Canada and the USA, where a new technique of Very Long Baseline Interferometry (VLBI) was developed.

Recognizing that the necessary radio links could not be extended to greater distances, the Canadian observers, who were first to make VLBI work, abandoned all idea of joining the two ends of an interferometer while the signals were actually being received. Instead, the signals were separately recorded on a tape recorder, of the type used in television recording because of its capacity for signals with wide bandwidths. The tapes were then transported to a laboratory where they were played back together as though the signals were being received together at the time of playback. For this system a local oscillator signal was still needed at each end of the interferometer; this was achieved without a radio link by using a pair of very stable oscillators which keep sufficiently closely in step even if separated by the length of the baseline. These were the atomic oscillators which are used in atomic clocks, containing caesium or rubidium resonators; later, hydrogen masers were used. Two identical oscillators had to be set up together, and then transported to the two telescope sites, which were often thousands of miles apart.

There were severe practical problems in the VLBI. It soon became possible to make interferometers which transcended national frontiers: the longest baseline in fact extended from the Crimea in USSR to California in USA. But to transport a working atomic oscillator on an aeroplane proved difficult – naturally the observers were asked what their mysterious packages contained. The answer 'a hydrogen maser' sounded like an explosive, while 'a caesium clock' elicited the fact that caesium is a poison which may not be transported by air. The word 'atomic' was equally difficult to use. So the local oscillators all became 'standard clocks', and the experimenters crossed the frontiers with less difficulty.

The results of the VLBI mainly concern the smallest of the quasars, which were not resolved by the radio-link interferometers. There are now known to be several quasars with angular diameters of about 10^{-3} arc seconds, which is the size of a tennis ball at a distance equal to the diameter of the earth. The best angular resolution attained by the VLBI was five times smaller again, at 2×10^{-4} arc seconds. This was achieved by using the baseline between the USA and the Crimea at 1·35-cm wavelength, which is the line wavelength of water vapour. One of the line sources, in the nebula W49, was still unresolved by this interferometer.

Unfilled Apertures

In a paper of 1952 M. Ryle discussed the theory and the possibilities of interferometers and showed that some advantages might be obtained from combining two aerials of different shapes into one interferometer. A small interferometer used for solar observations had already been constructed on this principle in Cambridge, and this was the first of a most interesting development which has no parallel in optical telescopes. The idea is illustrated in Figure 55.

All radio astronomers are in need of larger collecting areas than can be made, and all are in need of larger resolving powers.

The two do not exactly go together. Suppose the dimensions of (a) are sufficient for the required resolving power, say 10 minutes of arc, which would be very useful in radio-star work. In fact, to gain the advantages of an interferometer, two such apertures

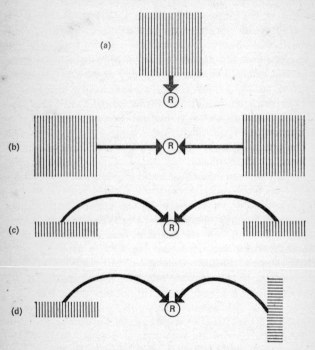

55. The development of the unfilled aperture. (a) Single aperture – pencil beam; (b) two apertures – interferometer; (c) interferometer with poor north–south resolution; (d) crossed interferometer

might be wanted, as in diagram (b). The interferometer shown in (c) has less collecting area, which is not always intolerable, but it has only a very low resolving power in the north–south direction. The interferometer in (d) with one aerial resolving in the east–west direction, and the other in the north–south direction, achieves the same resolution as the large pencil-beam aerial but with a great saving in area. For long wavelengths, which is just where angular resolution is so hard to achieve, the loss of sensitivity does not greatly matter, as nearly all celestial radio sources are stronger at long wavelengths.

An aerial of this type, in which only part is built while the resolution of the whole is obtained, is called an 'unfilled aperture'. B. Mills in Sydney built the first large instrument of this type. His radio telescope, known as the 'Mills Cross', is shown in Plate XIX. It is an interferometer, but the spacing between the two aerials has been reduced to zero, and the two arms are superimposed to form a cross. The advantage of the interferometer, in the improvement of contrast over a background, is retained, while various features of the background itself are still recorded. The essence of the Mills Cross lies in the achievement of a pencil-beam resolution at long wavelengths.

Aperture Synthesis

The economy of construction in the unfilled aperture, whether of the spaced aerial or of the Mills Cross type, does not entirely overcome the difficulties of obtaining high resolving power at radio wavelengths. There is still a lot of aerial to build, and, more seriously, there is still a lot of aerial to join together with transmission lines. Although the east–west arm can for most purposes be connected up once and for all, the north–south arm must be variable in its connections to allow for swinging the beam to different elevations. This is very hard to achieve.

Having made an unfilled aperture, the designer may well ask how much of an aperture must in fact be built. The answer is a surprising one: just two small sections only. But, as Michelson said for his one-dimensional interferometer, the spacings between these two must be changed in both size and direction, with a full series of observations made at each position. Figure 56 shows how this can be understood and used. The principle is called 'aperture synthesis', since a full-size aperture is obtained by adding up a series of small apertures.

In Figure 56, the synthesis of a single large aperture is shown. Dividing the aperture up into units of the size of the small move-

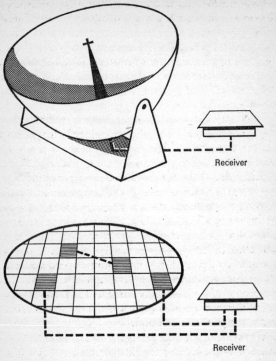

Receiver

Receiver

56. Aperture synthesis. A single large paraboloid may be simulated by two smaller aerials which can be placed at various distances apart in the same-sized aperture

able aerials, each possible contribution of angle and distance, such as AB, CD, etc., must be separately explored, moving an aerial between each observation and the next. It would seem that an impossibly long time would be taken for each sweep of the sky, but this is not so. Admittedly each series of observations must be completed before any part of a map can be drawn, but each series contains the information for constructing a map extending over a large region of the radio sky. If the same map were to be drawn from observations made with a conventional radio telescope with the same resolving power, that is to say a full aperture with a pencil beam, it would take almost the same time to cover the same extent of sky.

There is no need to go to the extreme of using two very small aerials; in fact this would bring serious troubles of sensitivity. A practical aerial of moderate cost must be sought between the two extreme possibilities of a fully filled aperture and a pair of moveable dipole aerials. The east–west aerial of a Mills Cross, or of a crossed-axis interferometer, is comparatively easy to build and needs no changing once it is built. If instead of the north–south arm we substitute a small moveable aerial, the same aperture can be 'synthesized' by moving this aerial along the north–south line. The actual construction of a map of the sky by this synthesis technique is a complicated matter. All the observations here are recorded on magnetic tape and fed into an electronic computer which then has to perform some millions of calculations. The computer then delivers a typed list of numbers, which can be used for drawing contour lines on a map. More conveniently, the computer itself can draw the map directly, on a plotting table connected directly to the computer output.

Aperture synthesis was first carried out successfully by J. H. Blythe, at Cambridge. The same principle is now applied in a series of radio telescopes at Cambridge, leading up to the notable 5-km telescope which uses eight parabolic reflectors, four of which are mobile and four fixed, all on an east–west line 5 km long. This telescope can produce a map of a small region of sky with an angular resolution of 2 arc seconds. There are also aperture synthesis arrays at work in Holland and in the USA (see Chapter 17). The latest proposal, for which construction has begun (in 1973) but which will not be complete until 1980, is for the Very Large Array, or VLA. This will be built in New Mexico for the National Radio Astronomy Observatory. It will consist of twenty-seven paraboloid reflectors spread along three railway lines radiating out from a central point. Each line will be 21 km long. The angular resolution at the shortest wavelength will be considerably better than 1 arc second, which is the size of star images as they appear on photographic plates from the best optical telescopes.

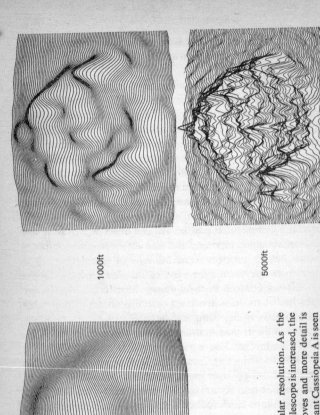

400ft

1000ft

5000ft

57. The effect of increasing angular resolution. As the baseline of an aperture-synthesis telescope is increased, the effective angular resolution improves and more detail is revealed. Here the supernova remnant Cassiopeia A is seen with synthesized apertures of 120 m, 300 m, and 1 500 m, at a wavelength of 6 cm (Cambridge 1-mile telescope)

The spectacular advance in resolution afforded by the technique of aperture synthesis is demonstrated in Figure 57. Here the radio source Cassiopeia A is drawn out using observations with the Cambridge 1-mile telescope. The three maps were obtained by using successively larger parts of the aperture. The first map is the best result that could be expected from a single paraboloid telescope.

RECEIVERS

In a radio telescope the part played by the radio receiver is similar to that of the photographic plate at the focusing point of an optical telescope. It is the point at which electromagnetic waves received over a large collecting area and focused to a point are converted into some recordable form. Descriptions of optical telescopes always refer to the size of the lenses or reflectors and to the arrangements for focusing the light. Not much mention is made of the photographic plate, and indeed the fact that new developments in recording involve replacing the photographic plate by an electronic image-converter seems to be of interest only to the astronomers using the telescopes. In the same way the receiving apparatus in a radio telescope, however novel, is hardly likely to be mentioned when a new radio telescope is described. The aerial to which it is attached will always steal the picture.

Nevertheless the receiving side of radio astronomy is of such importance that it cannot be overlooked; and furthermore there are some recent new developments in receiver techniques which will have as great an effect on radio astronomy as the image converter will have on optical astronomy.

The one outstanding requirement of a radio receiver used as part of a radio telescope is that it should be sensitive, and the history of these receivers is one of greater and greater emphasis on sensitivity. They are required to amplify and record the very small voltage that the aerial feeds into them, without distorting it by fluctuations in the receiver itself. From the modified wartime radar receivers used in the early days of radio astronomy there has been a steady progress in sensitivity and stability, and now the recent inventions of several different types of amplifier make possible an almost incredibly sensitive apparatus.

As the receiver's main task is to take a very small voltage and present it to a mechanical recorder as a larger one, the amplifying part of the set will be of great importance. Any amplifier tends to introduce its own unwanted signals into a receiver, whether it is a valve amplifier, a transistor amplifier, parametric amplifier, or a maser; the signals generated in any of these amplifiers have unfortunately many of the characteristics of the wanted signals. Both the wanted and the unwanted signals compounded of a random assortment of frequencies, are termed noise. Tuning in to cosmic noise on a receiver with loudspeaker output produces very much the same hissing sound as can be heard from an amplifier with no output at all, generated in the amplifier itself. Distinguishing between receiver noise and cosmic noise is thus one of the fundamental problems of a radio telescope receiver.

Amplifiers with a Low Noise Level

Cosmic radio noise can be specified as a temperature, and so can receiver noise. Figure 58 shows the noise temperature of a good receiver, using the best possible components, plotted against working frequency. The scales of this graph are plotted logarithmically, which gives a more even emphasis to the range of frequencies used in radio astronomy. Graph A shows the noise generated in valve amplifiers. For example, at a frequency of 300 MHz (wavelength of 1 m), a good triode valve generates only 300 K of noise, but at higher frequencies the situation becomes rapidly worse. Transistors have completely replaced triode valves, and their performance (Graph B) is still being improved. They are still beaten by crystal mixers (Graph C) at frequencies above 10 000 MHz.

The sky background itself adds a component to this noise temperature, of an amount which falls rapidly with frequency (Graph D), and at low frequencies, say less than 200 MHz; this is the main source of noise when a particular radio star is being examined against this background. No improvement in sensitivity can be made here by using better amplifiers. In fact 'the sky's the limit'. But at high frequencies the receiver is the limiting factor, and it is here that tremendous advances are being made. It is hoped that receiver noise at any frequency below 3 000 MHz may be held down below line E, corresponding to 20 K only, and giving improvements in sensitivity of over ten times that of conventional amplifiers.

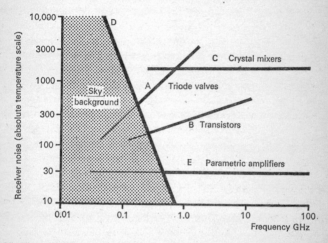

58. The variation of receiver noise with frequency for different kinds of amplifier

There are two different kinds of amplifier that can provide this amazing improvement: the maser and the parametric amplifier. We have already encountered maser amplification in the radio spectral line from the hydroxyl radical (page 127). The laboratory version is a similar quantum amplifier, but using a solid crystal rather than interstellar gas. A piece of crystalline material (a ruby has been used) held at a low temperature in a magnetic field can store away energy which it will emit as a radio frequency oscillation either strongly, when it is stimulated to do so, or more slowly if it is not stimulated. The stimulation comes from a radio wave of just the same frequency, so that a wave can be amplified by the fresh oscillation it sets off. The energy is stored in the crystal by another oscillator of a more conventional type and with a higher frequency, so that this energy pumped into the maser at one frequency emerges as an amplified signal at another.

A maser, then, is a crystal mounted in a waveguide and energized by an oscillator, so that radio waves from an aerial are amplified and fed into a normal receiver. The practical difficulties at present are great: the maser must be immersed in liquid helium, and it must be supplied with a precisely stabilized magnetic field and a precise amount of power from the 'pumping' oscillator. Nevertheless, it works. Most notably, the radio telescope at Onsala, in Sweden, has been equipped with a series of excellent masers constructed in the Chalmers Institute. Some of the best work on the 18-cm OH line has been done with a maser at Onsala.

The parametric amplifier is not a quantum device, nor does it need to operate at very low temperatures. The word 'parametric' implies that amplification is achieved by a change in the parameters of a circuit. Lord Rayleigh described a parametric amplification of a mechanical oscillation as long ago as 1883, and in fact the achievement of mechanical amplification dates much further back, to the unknown time when a child first found out how to 'work' himself up on a swing, by shifting his weight to produce a cyclic change of moment of inertia. The energy which is transferred to the swinging motion comes from this cyclic change of the main parameter of the system. Electrically, the parameter which is changed is usually the capacity of a condenser, and the change is induced by an oscillator, called a 'pump'. Energy from the pump goes into amplifying a radio signal, just as in an ordinary amplifier a battery or power pack supplies energy to an amplifying valve or transistor.

Parametric amplifiers are now used almost universally for receivers at frequencies between 300 MHz and 10 000 MHz. Most of them achieve a noise temperature in the range of 50 K to 100 K, but the very best ones achieve about 15 K. For this they must be refrigerated, and it is becoming common practice to use a refrigerator pump system which cools the heart of the amplifier system to a working temperature of 20 K or below.

Above 10 000 MHz, parametric amplifiers become very difficult to construct, and receivers fall back on one of the earliest radio techniques, the diode crystal mixer. Good mixers for frequencies near 100 GHz (100 000 MHz) are rare and valuable: the success of the molecular-line observations at Kitt Peak (Chapter 9) depended on a small batch of mixer diodes produced by Bell Telephone Laboratories, and not improved on or even equalled for several years.

The Computer as a Radio Receiver

No modern radio telescope seems to be complete without a computer, which is often working 'on-line', i.e. as part of the guiding and controlling system of the telescope. A control computer suitable for this task usually has a capability for performing several other operations almost simultaneously, and the observers have increasingly taken advantage of this spare capacity for such operations as taking averages of output signals and arranging for automatic tabulation, printing, and plotting. This has revealed new possibilities in data-processing, and experiments are now often planned in such a way that they are totally dependent on the computer.

An example of observations that depend entirely on the computer is provided by the searches for pulsars, carried out at Green Bank, Arecibo, and Jodrell Bank. The longest series of observations of this kind was carried out at Jodrell Bank by J. G. Davies and his colleagues, who set out to survey a wide strip of the Milky Way, using the Mark I radio telescope, whose beam covered about half a square degree. In this search over 4 000 beam areas were covered, each observation lasting about ten minutes. During each ten minutes the computer was required to record 16 000 numerical samples of the received signal, and search for any periodic signal contained in it which might have originated in a pulsar. During each ten minutes of observation the computer carried out the analysis of the previous 16 000 samples, trying out over 10 000 different periods and printing out any which contained a suspect pulsar signal. Even in one beam area this would be beyond human capability, but the computer can keep up the task indefinitely. During the search eighteen new pulsars were discovered, despite the fact that the same area of sky had already been searched by large radio telescopes. The 250-ft radio telescope was in fact the smallest radio telescope which had been used for discovering pulsars, but Davies's application of the computer enabled it to discover more pulsars than any telescope in the northern hemisphere. His rate of success is only beaten by the Molonglo radio telescope, which has unrivalled access to the whole of the southern sky.

Digital Spectroscopy

When optical astronomers observe the spectrum of the light from a star they usually split the light into its components by using a prism or a diffraction grating. This is impossible in radio astronomy; instead a noisy radio signal is to be analysed by electrical circuits into its various frequency components. The signal might for example contain a concentration of components near the frequency of 1 420·402 MHz, which is the natural frequency of the hydrogen line, and these components must be distinguished from the background of signal covering a wide band of frequencies around this line frequency.

Radio spectroscopy of this kind can be achieved by passing the signal through a bank of filters, each accepting a narrow range of frequencies, and recording all the separate outputs. This effective but clumsy method is now superseded by a process known as 'digital auto-correlation spectroscopy', originally introduced by S. Weinreb at Green Bank, USA. In this new method the signal is again converted into a series of digits, by very rapid sampling at the input of the computer. The presence of a line now manifests itself as a relationship between successive digits in this sequence, and the task of the computer is to search for and measure this 'auto-correlation'. This proves to be a very powerful method, with greater capabilities than the more obvious filter system. The spectra of Figures 30, 31, and 32 were obtained as direct computer outputs from this digital process.

Achievements in Sensitivity

Broadcast transmitters commonly attain powers of 10 kW or 100 kW, but such a small fraction of this is picked up by a normal receiver that its sensitivity has to reach down to the level of 10^{-12} watts, or one billionth of a watt. This achievement is nothing to the radio astronomer. He takes a receiver as sensitive as he can find, which may produce a noise level of only 10^{-15} watts, and he then proceeds to look for signals 10^{-4} less than this level. This is achieved by a process of averaging, by which the average output of the receiver over several seconds of time is measured. The longer the time, the more accurate the average. Also for a noise signal, the wider the frequency response of a receiver the more accurate is the measured average, and hence the greater becomes the sensitivity. Three ingredients go into the recipe for great sensitivity: a low-noise amplifier, a wide bandwidth, and a long averaging time.

Before the advent of low-noise amplifiers, some remarkable improvements were made by using the second ingredient, a wide bandwidth. Normal receivers may have bandwidths up to about 10 MHz but a travelling-wave tube (TWT) amplifier may have more than 1 000 MHz bandwidth. These amplifiers are used in microwave radio-communication links, where their large bandwidth allows many television or telephone signals to be transmitted simultaneously. Using a TWT receiver a radio telescope at Green Bank was able to detect a change in aerial temperature of one-hundredth of a degree; a direct measurement of this sensitivity can actually be made by using it to measure the temperature of a water bath surrounding a dummy aerial. A few drops of hot water in a large cold bath can make a detectable rise in temperature.

The smallest thermal signal detected directly in a single radio telescope is the thermal radiation from the most distant planets. Uranus and Neptune have been detected in this way as signals of only a few thousandths of a degree. With this sensitivity a radio telescope should be able to detect the thermal radio waves from an artificial satellite 300 km above the earth, even though the satellite temperature might be no higher than the freezing point of water.

Measuring even smaller signals depends on using a larger collecting area in the telescope, or on increasing the time over which the signal can be averaged. Both these factors operate in the aperture synthesis telescopes described earlier in this chapter, since in these instruments the areas of several separate telescopes are combined together, and the averaging process goes on for many hours or even days. The unit of signal strength or 'flux density' appropriate for catalogues of radio galaxies and quasars, is the Jansky; this unit equals 10^{-26} watts per square metre per unit bandwidth. The signals now being recorded by such telescopes as the Westerbork array and the 5-km telescope at Cambridge are measured in a unit one thousand times smaller, the milli-Jansky. If the receiver used in these telescopes were to collect radiation of one milli-Jansky over the whole area of our planet, using the same bandwidth, the total power would still be only 10^{-8} watts. A transmitter on the moon giving a power of less than 10^{-3} watts could give this minimum detectable signal. But such are the vastnesses of space that on the nearest star the most powerful transmitter yet constructed by man would be over a million times too feeble to make itself heard in our solar system.

AMATEUR RADIO ASTRONOMY

Grote Reber became a radio astronomer when he found that exploration of this world by amateur radio had been completed. As he put it:

My interest in radio astronomy began after reading the original articles by Karl Jansky. For some years previous I had been an ardent radio amateur and considerable of a DX addict, holding the call sign W9GFZ. After contacting over sixty countries and making W.A.C. [worked all continents], there did not appear to be any more worlds to conquer.

Any radio amateur who today feels the urge to explore further than his own planet need not step so far into the unknown as did Reber when he built his backyard radio telescope in 1937. Recording the signals from artificial satellites is an exciting way to start, and has proved valuable in practical teaching in schools. A notable contribution to space research has been made by Kettering Grammar School, which has so consistently monitored the telemetry signals from satellites that it has been able to announce the successful launch of several satellites from the Soviet Union ahead of the official announcement.

An amateur radio telescope capable of recording the signals from the sun, or from the two strongest discrete sources, Cygnus A and Cassiopeia A, might well be constructed at a school or technical college under the guidance of an experienced radio constructor. The variability of the solar radio waves makes an interesting astronomical study, and the ionospheric scintillation of the signals from the discrete sources might make another instructive subject. There is, however, some considerable difficulty in constructing sufficiently sensitive and stable receiving apparatus, and such a project ought not to be lightly undertaken.

Figure 59 gives a schematic idea of a complete radio telescope

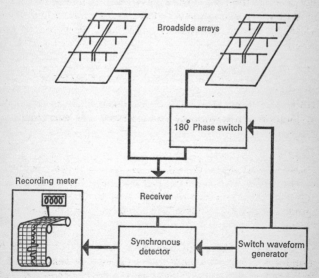

59. A simple interferometer radio telescope, suitable for recording radio waves from the sun

suitable for recording solar radio waves. The interferometer aerial uses broadside arrays of dipoles mounted over a reflecting wire sheet. From each aerial a coaxial transmission line runs back to the receiver, and there they join a 'phase-switching' unit which reverses the connection between them, alternating at a switching frequency of some hundreds of cycles per second. The switch is operated by crystal diodes. Then comes the receiver itself, followed by a special detector, which accepts only that part of the receiver output which alternates at the switch frequency. Finally this detector actuates via another amplifier a recording meter.

The most expensive single item is the recording meter, which records the output voltage in the form of an ink line on a long roll of paper. Provided with this, with ingenuity, and with access to the second-hand radio market, the amateur should have no trouble in making his own radio telescope. Before he sets out, however, he would be well advised to profit from the experience of others by joining an amateur group such as that of the British Astronomical Association.

CHAPTER 17

The World's Radio-Astronomical Observatories

EVERY three years the International Astronomical Union meets in General Assembly, and astronomers from all over the world join in discussions of every aspect of astronomy, from time-keeping to telescope making, and from the insides of stars to the outer parts of the universe. During recent assemblies the pressure on meetings on radio astronomy has been so great that separate symposia have been held for this subject alone, and it has even been necessary to restrict the subject matter further, say to 'The Radio Sun' or 'Radio Galaxies'. The last full symposium covering the whole of radio astronomy was in Paris in 1958, and it lasted a week. Only a few representatives of each radio observatory were there, and yet there were over two hundred attending. Twelve years earlier there were only about a dozen radio astronomers in the world, and twelve years before that only one. Where have they come from, where do they work, and what do they all do?

Astronomical research was until quite recently an individual enterprise, fostered by universities and receiving little assistance from the state. It is true that the Royal Greenwich Observatory was founded in 1675 by Charles II for the specific purpose of aiding navigation at sea, and it is true that this famous observatory, now removed to Herstmonceux, was run by the Admiralty until 1965, when it was taken over by the Science Research Council. But Flamsteed, the first Astronomer Royal, himself had to spend large sums of his own money in equipping the observatory, and state support through the Admiralty has until recently been restricted to maintaining a rather old-fashioned collection of instruments. Of the telescopes at Herstmonceux half date back to last century, and even in 1965 no telescope in the whole country had an aperture greater than 36 in. There is some encouragement in the new 98-in Newton telescope at Herstmonceux, but neither the Royal Greenwich Observatory nor the universities have fared well if one makes a comparison with the times of greatness brought to this country by Herschel, and to Ireland by Lord Rosse, who constructed and successfully used a 72-in reflector telescope in 1845. Again a contrast may be made with the endowment of the great telescopes in America, culminating with the $6 million contributed by the Rockefeller Foundation to the construction of the 200-in telescope of Mt Palomar.

Radio astronomy is developing in an age of extraordinarily rapid scientific progress. Since the war it has covered many of the phases of development which took ten times as long for optical astronomy. Starting with individual efforts, in which we ought properly to include those of Edison and Lodge, and other unsuccessful attempts to detect solar radio waves, we find the first radio astronomy pursued with great vigour but in isolation by Jansky and by Reber. It was only after the war that the stimulation of Hey's work in the Army Operational Research Group led to the beginnings of three established radio observatories which remain today as three of the largest in the world. Two, at Cambridge (the Mullard Observatory) and Manchester (Jodrell Bank), are university observatories. The third, in Sydney, is run by the Commonwealth Scientific and Industrial Research Organization as part of its Radiophysics Laboratory. The Sydney radio astronomers are therefore civil servants, just as they would be if they were in the Royal Greenwich Observatory, although in geographical location and in academic affinities they are very close to university research. They now work in parallel with a strong team from Sydney University under Mills, who built the large and important Mills Cross radio telescope near Sydney.

On the continent of Europe, it has been the long-established observatories of Meudon, in Paris, and of Leiden, that contributed most to the early days of radio astronomy. In these observatories, and particularly in Leiden, traditionally optical astronomers were brought into the new subject. Leiden has now joined with Utrecht and Gröningen in the Netherlands Foundation for Radio Astronomy, enabling a joint effort to be made in building the large synthesis radio telescope at Westerbork. Germany was later in making a major contribution, with the establishment of a new radio observatory at Bonn as part of the

Max Planck Institute; this observatory depends mainly on the 100-m reflector at Eifelsberg which began serious observations only in 1972.

In the USA the main lines of development came originally through the universities, of which the most notable was Harvard, where the hydrogen line was first detected. A major change of direction came with the establishment of the National Radio Astronomy Observatory (NRAO) at Green Bank, which was started by a consortium of several universities and became a national facility available to all universities. The major part of funds for radio astronomy in the USA now goes to NRAO, rather than to the individual universities.

The organization of astronomy in the USSR follows a different pattern, with the observatories and research institutes developing independently and without such direct links to the universities. Radio astronomy is based mainly on the two old-established optical observatories, Pulkova near Leningrad, and the Sternberg Institute in Moscow.

During the two decades 1950–70 there was a feeling of intense national competition in radio astronomy. This was partly due to the emergence of Soviet science, which provided a special stimulus to America, but it was felt at a more domestic level in the rivalry between Australian and British radio astronomers over the exciting discoveries in cosmology (Chapter 12). In the second of these decades a new pattern emerged which transcended these rivalries, and marked the beginning of a truly international community of radio astronomers. At this time radio astronomy was rapidly becoming 'big science', with large funds being channelled into a few observatories. The chance for more universities to start their own radio observatories was now past, and the only way ahead was collaboration. Several of the best observers therefore started a peripatetic existence, visiting the established observatories for their own chosen work. An important group started within the University of Maryland, USA, which relied entirely on observing at other institutions such as NRAO. The invention of the long-baseline interferometer (VLBI) then completed the process; observatories as far apart as the Crimea and California were joining in simultaneous observations, and the results were published under joint author-ship from several different countries.

The international cooperation that has now been achieved is well illustrated by the story of the radio source known as Cygnus X-3. This was originally discovered as an X-ray source, and later as a weak but variable radio source. A joint programme of obser-vations was undertaken by observers in the USA and Canada, monitoring the strength of the source every few days. One Saturday night in 1972 one of the Canadian observers, P. Gregory, was astonished to find that the radio signal from Cygnus X-3 had increased almost one thousand times, so that it had become one of the brightest objects in the sky. The Canadian observation was on the short wavelength of 3 cm, and it was urgent to look at other wavelengths and to follow the course of this unprece-dented outburst.

Within a few hours the result had been confirmed at Green Bank (NRAO), and the international telephone and telex system started humming with activity. By Monday all the major observa-tories were busily adapting their apparatus and reorganizing their programmes, and new results started to pour in. There was no reserve in sharing the information, and a full analysis could soon be made by Gregory and his colleagues from results sent in from all over the world. Even at this stage of collaboration it would not have been unreasonable for the various results to have been published in the various countries of their origin; instead all fifteen of them were collected together and published simul-taneously in one issue of the international journal *Nature*.

It should not, of course, be concluded from this story of open collaboration that personal and national ambitions are now entirely absent from the community of radio astronomers. There is nevertheless a marked contrast to some early attitudes, which are typified by the following anecdote. Sputnik III, the third man-made satellite, contained cosmic-ray detectors which were well suited to the detection of the energetic particles contained in trapped orbits in the earth's magnetic field. These radiation belts were in fact discovered not by Sputnik III, but over a year later by an American Explorer satellite; consequently they are named after J. van Allen, who devised the detection apparatus in that satellite. The van Allen belts were very nearly discovered by the space researchers in the USSR, and indeed they would have been apart from the circumstance that the Sputnik orbit only intersected the belts at the high point of the orbit, which occurred not over the USSR but over other parts of the world, and notably over Australia. The effect of the particles could not therefore be properly recorded in the USSR. The satellite signals were, however, recorded in Australia, in a code only known to the Soviet workers. If either the recordings had been sent to Moscow, or if the details of the code had been revealed to Sydney, the radiation belts would have been discovered immediately. But neither of these forms of cooperation was thought possible at that time of competition and political tension, and the discovery was postponed until the Explorer was launched. There was no ultimate loss to science, but there was certainly a loss to Australia and the USSR, and some good fortune for van Allen.

Such incidents are luckily a thing of the past, even in the other delicate fields of space research, and it is now a commonplace to find, for example, Jodrell Bank tracking a Soviet spacecraft and sending the telemetry recordings direct to the President of the USSR Academy of Science in Moscow, or to find teams of radio astronomers from the USA working in the Crimea for a VLBI experiment.

A List of Radio Observatories

The following list of observatories is not intended to be exhaustive and complete. It includes the major radio observatories of each country and shows how each has made and is now making its own special contributions to astronomy. Emphasis on some particular discoveries, or on particular instruments, may often mean only that they have particularly impressed or interested the author.

As he glances through this list, each reader will be looking for some aspect of the observatories which particularly interests him, whether it be their history, their organization, or perhaps only the relative dimensions of their various radio telescopes. Whatever this special interest he should bear in mind primarily the differentiation in objectives that is now so evident in the various observatories. There are small observatories, for example, whose object is a simple but effective 'solar patrol', a continuous watch on the sun's radio emission, possibly only at one radio frequency. Research of this limited nature is well suited to smaller universities, where little effort can be spared for post-graduate training, but where a critical approach to experimental work and to the handling of observational results can only effectively be taught by practice.

There are also government establishments, such as the Naval Research Laboratory in Washington, and the Radiophysics Laboratory in Sydney, where national resources are used to ensure on the one hand the continuance of a healthy tradition of scientific research, and on the other a reserve of scientific manpower. In these the government has clearly recognized the value of encouraging individual scientific enterprise, and many diverse lines of research may be concentrated in one very active establishment.

Again, old-established observatories are now providing a foundation for radio observatories, and these are likely to follow in some way the traditions of the optical work; as for instance Leiden Observatory has guided Dutch radio astronomy primarily towards the determination of galactic structure by 21-cm hydrogen-line observations. All these influences are reflected in the size of the observatories, in the kind of work they undertake, and more particularly in the radio telescopes they construct.

The list which follows is in alphabetical order of countries.

AUSTRALIA

The Radiophysics Laboratory, Sydney

This laboratory is part of the Commonwealth Scientific and Industrial Research Organization. During the war it was a centre for radar research. Immediately at the end of the war, work was begun there on solar radio waves under the direction of J. L. Pawsey. With a record of rapid and most enterprising development this laboratory now stands as one of the longest established and most distinguished of the world's radio observatories. To record all its outstanding contributions would be to repeat much of this book, since they cover most fields of interest.

Pawsey's solar work was achieved with the cliff interferometer, using a single aerial mounted on top of the 300-ft cliff at Dover Heights, just south of the entrance to Sydney Harbour. It was here also that John Bolton made the first interferometer observations of a 'radio star', and with the help of another cliff interferometer mounted in New Zealand found the first positions of several of these then unidentified objects. The first identifications are also to be credited to Bolton, who suggested that the Crab nebula and the nebula NGC 5128 in Centaurus were two of the discrete sources which he had found.

The solar observations started by Pawsey were developed by J. P. Wild in two ways. He realized that recordings of solar radio waves on a fixed frequency were totally inadequate for understanding the complex bursts of emission from sunspots and flares, and he developed a radio spectrograph which could display a wide sweep of the radio spectrum on film several times a second. The complete picture of the structure of bursts obtained by Wild's radio spectrograph has achieved far more than a mere classification of different types of bursts: it gave the first demonstration of the rapidly moving disturbances which travel out through the solar corona at speeds up to a tenth the speed of light.

Wild's second contribution to solar studies is called a radio heliograph. This instrument receives solar radio waves at a fixed wavelength, but it produces a map of the solar surface with angular resolution of 5 arc minutes, displayed on a cathode ray tube so that it can be filmed to provide a dynamic record of the development and movements of the radio emission from solar flares. Wild's radio heliograph is at Culgoora, 400 km north of Sydney. It consists of ninety-six small paraboloid aerials arranged on a circle 2 km in diameter.

The Radiophysics Laboratory has a fine 210-ft reflector telescope, located at Parkes, near Sydney. This was used notably for the occultation measurements on 3C 273 (Chapter 11); its unique location in the southern hemisphere has also made it invaluable for all kinds of radio astronomy, particularly for galactic spectral line work and for extragalactic radio sources.

Sydney University

The Departments of Electrical Engineering and of Physics now contribute separately to Australian radio astronomy through two former members of the Radiophysics Laboratory, Professor W. H. Christiansen and Professor B. Y. Mills. Both were involved in the early use of unfilled-aperture radio telescopes; two of these instruments, in the form of crossed arrays, have become known as the Chris-Cross and the Mills Cross. The largest version of the Mills Cross has been responsible for the discovery of the majority of the pulsars in the southern hemisphere.

Hobart, Tasmania

In this university a remarkable partnership between the veteran Grote Reber and the ionospheric physicist G. R. Ellis has led to very fruitful observations in the frequency range 2 MHz to 20 MHz, especially of the planet Jupiter. For this work an array 2 000 ft square has been constructed.

CANADA

Distinguished work was done in the early days of radio astronomy by D. W. R. McKinley and P. M. Millman, who solved many of the problems of meteor radar echoes, and by A. E. Covington, whose measurements of 10-cm-wavelength solar radio waves now present an unbroken sequence over two solar cycles.

There are now two major observatories in Canada, both run by the National Research Council, and both supporting research within the council's own laboratories as well as in the universities. The major radio telescope in Canada is the 150-ft paraboloid in Algonquin National Park, near Ottawa. The western observatory is at Penticton, in the Rocky Mountains; this is a most attractive site which combines great natural beauty with a very low level of radio interference. Large Mills Cross telescopes, for 22 MHz and 10 MHz, are used for measuring the spectra of discrete sources at the lowest possible frequencies, and an 85-ft paraboloid has been used notably for VLBI, pulsar, and hydrogen-line studies.

GREAT BRITAIN

Cambridge

Radio astronomy in the Cavendish Laboratory grew out of a tradition of ionospheric research, under J. A. Ratcliffe. Stimulus came from Hey's discoveries during the war, and from wartime improvements in radio techniques. The whole development has been directed by M. Ryle, now Sir Martin Ryle, Astronomer Royal. Starting on solar radio waves, his research group developed the techniques of interferometry for measuring the radio-brightness distribution over the sun, and went on to apply these techniques to other radio sources. Early successes included the accurate positions of Cygnus A and Cassiopeia A, leading to their identifications, and the measurement of their angular diameters.

Developments of interferometry at Cambridge led eventually to the aperture-synthesis telescopes which are now the main instruments at the radio observatory. The two largest are known as the '1-mile' and the '5-km' telescopes. The latter telescope, completed in 1972, produces maps of radio sources with an angular resolution of one arc second, which is the size of the images on the best photographs taken with large optical telescopes. With comparable angular resolution the human eye could read ordinary newsprint at a distance of 300 m.

Other notable advances at Cambridge include the discoveries of interplanetary and interstellar scintillation, and the discovery of pulsars. The main theme has been the study of extragalactic radio sources with a succession of original ideas incorporated in new types of radio telescopes. The funds for the research are now largely provided by the Science Research Council, but a major stage of development was assisted by a benefaction received by Cambridge University from Mullard Ltd; the observatory is consequently known as the Mullard Radio Astronomy Observatory.

Jodrell Bank

This world-famous observatory has grown from the meteor radar experiments conducted in 1947 by its director, now Professor Sir Bernard Lovell, in a trailer on the site of the present buildings. The 250-ft paraboloid, completed in 1957 and known as Mark I, is famed throughout the world through its occasional use in tracking deep-space probes, especially in the early days of space research when the USA and the USSR had no comparable means of communicating with space vehicles. It has recently been partly rebuilt, with an improved reflector surface and drive system, so that despite its age it is still in the forefront of research.

Following a tradition of the University of Manchester, Jodrell Bank has pioneered the application of digital computers to telescope control and data analysis, as for example in the discovery by J. G. Davies of the majority of the pulsars in the northern sky.

The outstanding astronomical contribution has been the discovery of the small-diameter radio sources, later known as the quasars, in the use by H. P. Palmer of long-baseline interferometers. As a by-product, a technique devised at Jodrell Bank by R. Hanbury Brown and R. Q. Twiss for measuring radio-source diameters has now been applied by Hanbury Brown to visible stars; his optical interferometer at Narrabri, in Australia, has provided a classic and outstanding advance on Michelson's original measurements.

Jodrell Bank has also pioneered the technique of radar aperture synthesis, which provided the radar maps of the moon, and the first studies of radio emission from flare stars. It is interesting also to recall the construction of a fixed 220-ft paraboloid in 1950, intended for radar studies of cosmic ray showers. The paraboloid was in fact used instead for pioneering work in galactic and extragalactic astronomy, while the detection of cosmic-ray showers was deferred until fifteen years later, when it was achieved in the first detection of a radio pulse directly transmitted by the shower itself.

Jodrell Bank is part of the Physics Department of the University of Manchester. It is officially known as the Nuffield Radio Astronomy Laboratories, in commemoration of a benefaction from the Nuffield Foundation. It operates an out-station at Defford, near Gt Malvern. Here there is an 85-ft reflector, used with the Mark I telescope for long-baseline interferometry. Until 1972, this reflector was one of a pair used by the Royal Radar Establishment for accurate position-finding on discrete radio sources.

FRANCE

The Paris Observatory at Meudon has a long tradition of solar radio observations, originally under the leadership of J. F. Denisse. Radio observations of many kinds are now made at an observatory at Nançay, south of the Loire. The first instrument there was a 32-element interferometer working at 2-m wavelength for locating radio outbursts on the sun. Much of the work of this observatory was originally concerned with radio emission from the sun, and with solar–terrestrial relationships, but the construction of a major reflecting telescope turned most of the research towards galactic line work, and extragalactic sources.

The Nançay radio telescope is a reflector of the Kraus type, measuring 300 m by 30 m. The surface is sufficiently accurate for 6-cm wavelength, making it one of the most powerful radio telescopes in the world.

Another research group at Bordeaux is developing new observational techniques for millimetre-wave radio astronomy. As at Nançay, the initial observations are solar, but the intention is to develop interferometric methods for discrete radio sources of all types.

GERMANY

Until a few years ago, Germany's chief contribution to radio astronomy was the network of 'Wurzburg' radar stations established along the coast of Europe and abandoned after the war. Each station was equipped with an 8-m paraboloid aerial of excellent quality, and many of these aerials were later appropriated by new radio observatories both in Europe and America. They were used, for example, in an interferometer at Cambridge which found accurate positions for Cygnus A and Cassiopeia A. A major observatory has now been built near Bonn by the recently founded Max Planck Institute for Radio Astronomy. This observatory has a 100-m steerable telescope, the largest in the world. The surface is sufficiently accurate for 3-cm wavelength. First observations with this telescope were made in 1972.

INDIA

The Tata Institute in Bombay has enabled Professor Swarup, who had previously worked in Australia and the USA, to construct a major radio telescope designed specifically for recording lunar occultations. It is a large cylindrical paraboloid, aligned north–south on steeply sloping ground so that the axis is parallel to the rotation axis of the earth. This is effectively the 'equatorial mount' familiar in optical astronomy, and it is very convenient for following the moon's motion across the sky. This remarkable instrument is at Ootacamund, a hill-station in the south of India, usually referred to briefly as 'Ooty'.

ITALY

The Institute of Physics at Bologna has a spectacular radio telescope of the Mills Cross type, known as the 'Northern Cross'. Each arm measures 600 m by 30 m. It is used for surveys of discrete radio sources at 72-cm wavelength; the Bologna catalogues of fluxes and positions of many thousands of sources have made an important contribution to identifications of extragalactic sources and to studies of their spectra.

JAPAN

Almost all the work of radio observatories in Japan has been concerned with the sun, and most of it has been carried out with new interferometric techniques at short wavelengths. The observatory at Nagoya, under H. Tanaka, has become the leading observatory in the world for regular daily measurements of solar radio flux. At Nagoya there are also some remarkable and complex interferometers which incorporate polarimeters; these produce maps of the sources of polarized radio emission on the face of the sun.

NETHERLANDS

A government foundation, the Netherlands Foundation for Radio Astronomy, draws together the resources of the observatories at Leiden, Utrecht, and Gröningen. The great strength of the foundation lies in its direct association with a long tradition of optical astronomy. It was at Leiden that J. H. Oort prompted H. C. van de Hulst in his proposal to use the 21-cm hydrogen line for studying galactic structure. Oort's work on the Crab nebula also provided a link between optical and radio astronomy through his recognition that synchrotron emission covered the whole observable spectrum of radiation from the nebula.

There are two observatories, at Dwingeloo and at Westerbork. The major instrument at Westerbork is an aperture-synthesis array of twelve parabolic reflectors, giving the resolution of a 1·6 km aperture at wavelengths of 6 cm and 21 cm. This has been spectacularly successful in producing detailed maps of the radio brightness distribution across extragalactic nebulae.

USA

Although for many years radio astronomy developed slowly in the USA, despite the pioneering efforts of Jansky and Reber, developments since the 1950s have been widespread and fruitful. Most of the growth took place originally in universities, and many of these now support observatories of various kinds. The Carnegie Institution and the Naval Research Laboratory were also early in the field. But it has now been clearly recognized in America that the largest and most expensive pieces of equipment, which may be essential for advances in any science, cannot be part of the normal equipment of every university. Observatory work is therefore concentrated in the National Radio Astronomy Observatory (NRAO), at Green Bank in West Virginia, which caters for visiting observers from all universities.

NRAO has several major radio telescopes at Green Bank, and in addition a millimetre-wave telescope at Kitt Peak in Arizona. It is this 36-ft telescope that made so many spectacular discoveries in molecular spectral lines in 1970 and 1971. At Green Bank there is a 300-ft transit telescope, a 140-ft fully steerable telescope, and an interferometer used for aperture synthesis. A major development will be the construction, over an eight-year period ending in 1980, of a new telescope called VLA, standing for 'very large array'. This will be an aperture-synthesis telescope, using twenty-seven reflectors each 25 m in diameter, mounted on three railway tracks each extending 21 km from a central point. This telescope should give an angular resolution better than 0·5 arc seconds.

A notable centre for radio astronomy is at the University of Maryland, where there are no observing facilities at all, and there is instead a complete reliance on visits to Green Bank and other observatories. Other universities have important telescopes of their own: for example, Ohio State has a large Kraus reflector, Illinois has a large parabolic trough, Michigan has an 85-ft reflector, all of which have made notable contributions to astronomy. Harvard and the Massachusetts Institute of Technology (MIT) are important centres: the 21-cm line was discovered at Harvard by H. I. Ewen, and MIT has been involved in VLBI observations, using in particular the 120-ft reflector known as 'Haystack'. This reflector was originally built mainly for lunar radar, and it has produced some of the most detailed radar maps of the lunar surface.

The California Institute of Technology, which shares in the running of the large optical telescopes in California, has an interferometer consisting of a fixed 150-ft paraboloid and two 90-ft paraboloids on railway tracks. This instrument was built by J. G. Bolton, who was a pioneer of radio astronomy in Sydney.

The largest reflector telescope in the world is the 1 000-ft spherical reflector at Arecibo, Puerto Rico. This was built by Cornell University for planetary radar and for ionospheric studies, but increasingly it has been applied to other branches of astronomy, notably to pulsars, where the large collecting area has provided a very useful improvement in sensitivity. Cornell

University has established a special liaison with the University of Sydney in Australia, and there are frequent interchanges of observers and apparatus.

USSR

The resurgence of astronomy after the terrible destruction that came with the war was accompanied by a rapid growth of radio astronomy. The main contribution of the USSR has undoubtedly been the theoretical work of the Sternberg Institute in Moscow, where I. S. Shklovsky and V. L. Ginzburg first applied the theory of synchrotron emission to galactic and extragalactic sources. Major radio telescopes are now in use at several centres.

At Pulkova Observatory, famous for its optical work, there is a reflector constructed as a sector of a paraboloid, 100 m long and 3 m wide, and working at 8-mm wavelength. A larger version is under construction in the Caucasus, under the direction of Y. Pariisky.

The Lebedev Institute of Moscow has an out-station at Serpukhov, where there is a large Mills Cross aerial suitable for use at long wavelengths. There is also a very fine 22-m reflector telescope suitable for millimetre wavelengths. The surface of this reflector is accurate to about 0·5 mm, including gravitational deflections. A similar reflector in the Crimea was used for the intercontinental VLBI in a cooperative experiment with America.

Kharkov, in the Ukraine, has a large low-frequency array used for extending the spectra of discrete sources to 10 MHz. This array is an interferometer system in the form of a huge 'T'; it comprises 2 000 dipoles and covers a total area of $37\frac{1}{2}$ acres.

WHAT NEXT?

It is now increasingly evident that the major part of radio astronomy must be concentrated into the few large institutions that can attract sufficient funds for large modern instruments. This is a sad situation for many of the smaller observatories, but the example of the National Radio Astronomy Observatory in the USA should encourage them to join the large cooperative efforts that are becoming the order of the day. There is little useful to be done with a reflector telescope unless it is very large, like the 100-m telescope at Bonn, or very accurate, like the 36-ft millimetre-wave telescope at Kitt Peak; these are expensive instruments. Aperture-synthesis telescopes for resolutions of the order of 1 arc second, and long-baseline interferometry for resolutions down to 0·001 seconds, have a great future but are also becoming expensive in an era of constrained budgets for science.

Although there will be an increasing concentration into a few major observatories, the examples of Japan in millimetre-wave solar astronomy and of India in the lunar occultation radio telescope at Ooty, both show the usefulness of well-directed and selective effort. There need be no discouragement for new plans, such as those of the new Nicolaus Copernicus University in Torun, Poland, where a major effort will be directed towards a study of interplanetary scintillation.

The pattern of international cooperation is already familiar in space research. Here no university can expect to launch its own rockets and satellites unaided, and most countries rely on the hospitality of the USA or the USSR if they wish to participate in satellite experiments. In space research there has been great benefit from the interchange of personnel and ideas between the very flexible and mobile research groups that have participated, and there is every reason to expect similar benefits in the new pattern of radio astronomy.

Will there continue to be such rapid advances in radio astronomy as we saw during the 1960s and early 1970s? The only sure answer is to place oneself ten years back and see if prediction was possible then. Who could then have predicted the discovery of the pulsars, or the molecular spectral lines? Who could have predicted the complexities of the quasar red shifts? But in fact these surprises all arose from the steady hard work of the observatories, and the general lines of progress could dimly have been foreseen. Further, the time scale of telescope construction was already lengthening, and the lines of development of today's instruments was already becoming clear. We ought therefore to venture some predicting for 1980 and beyond.

The new generation of radio telescopes is already clearly seen as separating into three types. First there should be one or two more of the very large steerable parabolic reflectors, on the lines of the 100-m Bonn telescope; one such is the projected Jodrell Bank 115-m Mark V telescope to be built in Wales. Such large reflectors are not practicable for millimetre wavelengths, and a second category is emerging, with apertures up to 30 m in diameter and with surface accuracy approaching 0·1 mm. There is enormous scope for new instruments in this category, and there are formidable engineering problems to overcome. Finally there are the aperture-synthesis telescopes, such as the new Cambridge 5-km telescope and the Very Large Array (VLA) in New Mexico. Perhaps the main need here is to extend the sky coverage of these high-resolution telescopes to the southern hemisphere, for example by a new aperture-synthesis telescope in Australia.

Instruments and receivers already rely on digital techniques, and this reliance is rapidly increasing. Altogether there is a bewildering increase in complexity of receiving techniques, which makes observing a more expert business in which the traditional combination of expertise both in technology and astronomy becomes more difficult.

Astronomically, it is easy to see the fields that are growing. It is no longer so exciting cosmologically merely to count extragalactic radio sources: instead the physics of processes inside them are crying out for study through the new high-resolution maps of aperture-synthesis. It is no longer so exciting to discover a new spectral line as it is to relate two lines together and elucidate the conditions in collapsing stellar clouds. Pulsars are exciting, not so much as a spectacular source of radio pulses but as the seat of processes which we would do well to understand for the sake of our basic physics.

Above all, we now see the universe in the process of change. Within the galaxy we see changes in the OH spectral lines in periods of a few months, changes in pulsars in a range of periods from microseconds to their lifetimes of a million years or more, and we see spectacular changes in the X-ray source Cygnus X-3 and possibly in other stellar objects. Some extragalactic sources are showing changes within themselves over periods of only a few days or weeks, but the great physical insight which these sources afford us is not so much within themselves but on the largest scale – that of the universe itself. Our sight now penetrates beyond the local average of cosmological space and time, out to the distant regions where there is evidence of the overwhelming changes in the universe as it expands out from the primeval explosion, and as the galaxies form, explode as quasars and radio galaxies, and condense into the comparative peace of the Milky Way.

Finally, radio astronomy must now take its proper place among the many new branches of astronomy that have sprung up in the last few years. X-ray astronomy can now claim some hundreds of identified celestial objects, and gamma-ray astronomy is following close after. Satellite-borne telescopes will soon extend the infra-red and ultra-violet spectrum to complete the wide range of wavelengths available to the observer. The Crab nebula pulsar can be observed both at long radio wavelengths, and at the most energetic gamma-rays, covering a range of over forty octaves. Of these only ten are within the radio range. It would be short-sighted indeed to suggest that the spectacular discoveries in radio astronomy will not be matched or even overtaken by developing techniques which apply to those other thirty octaves.

Index

Roman numerals refer to numbered plates

Alfven, H., 83
ammonia: in space, 129
amplifiers: radio telescopes, 240–43
Andromeda nebula, 40, 135–6, 137, 138–9, I
aperture synthesis, 150, 186–8, 235–9, 257
Appleton, Sir E., 68, 69, 70, 212
arch prominence, XV
Arecibo, 194, 227, 262
Arend-Roland comet, 208, 216–18
asteroids, 36, 214
Australia: observatories, 250, 254–6

Baade, W., 49, 87, 143, 147
Babylon, 15
Barrett, A. H., 128
Bay, Z., 183–4
Bell, J., 95–6
Bennett comet, 217, XVI
Biela's comet, 210
big bang theory, 173–5
Blythe, J. H., 237
Bologna: observatories, 260
Bolton, J. G., 82, 142, 255, 262
Bondi, H., 166
Bonn: observatories, 250–51, 260
Brahe, Tycho, 16, 82, 88, 89, 220
Branson, N. J. B. A., 201
brightness: sun, 55–7
Burbidge, G., 83
Burke, B. F., 202
bursts, solar, 72–80

Cambridge: observatories, 257
Canada: observatories, 256–7
carbon monoxide: in space, 130
Cassiopeia A, 44, 86–91, 238, X

Centaurus A, 148–9
Christiansen, W. H., 116, 256
comets, 36, 208–10, 216–18, XVI
computers: radio telescopes, 243–4
'confusion': radio telescopes, 168–9
Conway, R. G., 84
corona, solar, 37, 52–67, XIII
cosmology, 51, 164–76
Covington, A. E., 256
Crab nebula, 32, 63–4, 81–5, 97, 98–9, 99–100, 101, VIII, IX
Cygnus A, 22–3, 48–50, 143–6, XI
Cygnus X-3, 93, 252

Davies, J. G., 214, 215, 243, 258
Denisse, J. F., 259
de Sitter, W., 164–6
deuterium, 126–7, 133
Dewhirst, D., 87
Dicke, R., 174–5
diffraction: of radio waves by solar corona, 65
digital spectroscopy, 244
diode crystal mixer, 243
Dombrovsky, V. A., 84
Doppler effect: hydrogen line, 116–20
Downes, D., 114
Dreyer, J. L. F., 48
dwarf stars, 96

eclipses: sun, 57–9
Eddington, Sir A. S., 47
Edison, T., 17–19
Eifelsberg 100-m reflector telescope, 251, XX
Einstein, A., 164–6
Ellis, G. R., 256
Elsmore, B., 180

267

equipartition, 144
Evans, J., 185
Ewen, H. I., 116, 262
extragalactic nebulae: *see* nebulae

Faraday effect, 121–3
flare stars, 91–3
flares, solar, 72–80
formaldehyde: in space, 130
Fowler, W., 162
France: observatories, 259
Franklin, K. L., 202
Fraunhofer lines, 16–17, 52
Fresnel diffraction theory, 213

galaxies, 38–43, 46–51, 135–55, 171, **III**. *See also* radio galaxies
Galileo, 16
Galt, J., 106
Germany: observatories, 250–51, 259–60
Giacobini's comet, 210, 212
Gibson, J. E., 181
Ginzburg, V. L., 84, 263
glitches, 99–100
Gold, T., 166
gravitational energy, 162
Great Britain: observatories, 249, 250, 257–9
Green Bank: observatories, 251, 261–2
Greenwich: observatories, 249
Gregory, P., 252

Hagen, J. P., 57, 58
Hale, G. E., 11, 46
Halley's comet, 208, 209
Hanbury Brown, R., 258
Heightman, D. W., 69
Herlofson, N., 83
Herschel, Sir W., 11
Hertz, H., 11, 17
Hewish, A., 94–6
Hey, J. S., 22, 48, 70, 211, 212
Hjellming, R. M., 91
Hindman, J. V., 116

Hobart: observatories, 256
Hoyle, F., 162, 166, 175, 176
Hubble, E., 39–40, 46, 47, 51, 139, 165
Hulst, H. C. van de, 115–16, 261
hydrogen: H II regions, 112–13; spectral line, 45, 113–20
hydroxyl, 127–9

IC 443, 88–9, 107, **XII**
India: observatories, 260
interferometry, 56, 150, 159–61, 228–39
interplanetary space, 37–8
interstellar gas, 108–24
ionosphere: effect of solar flares, 77–80
Italy: observatories, 260

Jansky (unit of measurement), 246
Jansky, K. G., 11, 20, 39, 68
Japan: observatories, 260–61
Jodrell Bank: Mark IA radio telescope, **XXI**; observatories, 258–9
Jupiter, 35, 194, 198, 199–206, **XVII**

Kennelly, A. E., 18–19
Kepler, J., 16, 82, 220

Landau, L., 96
Leiden: observatories, 250, 261
libration: moon, 184–5
life: origins, 125–6
light waves, 25. *See also* spectroscopy
Lockyer, Sir J. N., 17
Lodge, Sir O., 19
long-baseline interferometry, 231–3
Lovell, Sir A. C. B., 91, 212, 214, 215, 258
lunar: *see* moon
Lynds, C. R., 149
Lyne, A., 106

M 82, 149
McCullough, T. P., 197
McKinley, D. W. R., 214, 256
McVittie, G. C., 159
magnetic field: interstellar space, 121–3; pulsars, 100–102; sun, 71–4
magnetobremsstrahlung, 44
Mars, 194, 195, 198, **XXII**
maser amplification, 128, 241–2
Maxwell, A., 114
Maxwell, J. Clerk, 17
Mayer, C. H., 197
Mercury, 194, 195, 196, 198
Messier, C., 48, 82
meteors, 22, 36–7, 207–8, 211–16, **XVIII**
Meudon: observatories, 259
Michelson, A. A., 56, 229, 230
Milky Way, 28, 38–45, **VI**; synchrotron radiation, 108–12
Millman, P. M., 256
Mills Cross radio telescope, 235, 256, 256–7, **XIX**
Minkowski, R., 49, 87, 143, 146
Minnett, H. C., 181
molecules: in space, 125–34
moon, 30, 177–91
Mullard Observatory, 257

Nagoya: observatories, 260–61
Naismith, R., 212
Nançay: observatories, 259
nebulae, 39–40, 46–8, **I, II, V, VII, VIII**. *See also* Andromeda nebula, Cassiopeia A, Centaurus A, Crab nebula, IC 443
Neptune, 198
Netherlands: observatories, 250, 261
neutron stars, 96–8
Newton, Sir I., 16
non-thermal radiation, 31–2
North Galactic Spur, 110, 111
novae, 91–2

observatories, 249–66

occultations, radio, 63–5, 157, 179–80
Olbers, H. W. M., 165
Oort, J. H., 23, 115, 261
Orion nebula, 130, **VII**
outbursts, solar, 72–80

Pacini, F., 97
Palmer, H. P., 50, 159–60, 258
parabolic reflectors: radio telescopes, 221–8
parametric amplifiers, 242
Pariisky, Y., 263
Pawsey, J. L., 23, 231, 254–5
Penzias, A., 174
Perseus A, 147–8
Piddington, J. H., 181
planets, 35–6, 192–206
polarization: Faraday effect, 121–3; radiation from Crab nebula, 83–4, 85
Ponsonby, J. E. B., 187
pulsars, 29, 43, 84–5, 94–107, 122, 124, 243–4, **IX**

quasars, 50–51, 156–63, 233, **IV**

radar, 69–70, 183–8, 190–91, 193–6, 211–13
radiation: non-thermal, 31–2; thermal, 29–31, 196–9
radio galaxies, 48–50, 142–55, **III, XI**
radio heliograph, 255
radio stars, 43, 91–3
radio telescopes, 219–48, 254–63, 265, **XIX, XX, XXI**
radio waves: characteristics, 26–8
Radiophysics Laboratory, Sydney, 254–6
Ratcliffe, J. A., 257
Rayleigh, J. W. Strutt, 3rd Baron, 184, 242
Rayleigh–Jeans Law, 30
Reber, G., 11, 21, 48, 221, 246, 256
receivers: radio telescopes, 239–46

red shift: spectroscopy, 139–40, 141–2, 158–9, 166
refraction: of radio waves by solar corona, 63–5
relativistic beam compression, 100, 102
relict radiation, 173–5
Rickett, B. J., 124
Rosse, W. Parsons, 3rd Earl, 250
rotation curves, 137–9
Royal Greenwich Observatory, 249
Russia: observatories, 251, 263
Ryle, Sir M., 23, 168, 170–72, 229, 233, 257

Sagittarius A, 114
Sandage, A., 156
Saturn, 198
Schatzman, E., 91
Schmidt, M., 157
Sciama, D., 174
scintillation, 37–8, 65–7, 94–5, 106–7, 123, 124
Shklovsky, I. S., 84, 263
Sitter, W. de, 164–6
Slipher, V. M., 139
Sloanaker, R. M., 197
solar system: see moon; planets; sun
Southworth, G. C., 53
space: curvature, 175–6
space research, 35–6
spectroscopy, 16–17, 45, 52–3, 113–20, 126–33, 158–9, 244
steady-state theory, 166–7, 173
Stefan's Quintet, 140–42
Stewart, G. S., 211, 212
Stokes, G. G., 16–17
sun, 30–31, 33–4, 37, 52–67, XIII
sunspots, 59–60, 68–80, XIV
supernovae, 81–6, 88, 89, X
Swarup, G., 260
Sydney: observatories, 250, 254–6
synchrotron radiation, 44, 83–5, 108–12, 136–7

Tanaka, H., 260
telescopes, 46, 222. See also radio telescopes
temperature: H II regions, 113; moon, 180–83; of an emitter, 30–31; planets, 196–9; stars, 41; sun, 30–31, 52–5, 60–61
thermal radiation, 29–31, 196–9
Thomson, J. H., 187
3C 147, 152–3
3C 295, 146–7
Twiss, R. Q., 258

unfilled apertures: radio telescopes, 233–5
United States: observatories, 251, 261–3
universe: cosmology, 51, 164–76
Uranus, 198
U.S.S.R.: observatories, 251, 263

Van Allen belts, 252–3
Van de Hulst, H., 115–16
Vashakidze, M. A., 84
Vela pulsar, 98, 99
Venus, 194, 195, 196, 198, 199
Virgo A, 149, III
VLBI, 160–61, 232–3
Vonberg, D. D., 229

Wade, C. M., 91
water vapour: in space, 129, 130
Weinreb, S., 244
Westerbork: observatories, 250, 261
Whipple, F. L., 211, 213
Whirlpool nebula, II
white dwarfs, 96
Whitfield, G. R., 217
Wild, J. P., 77, 255
Wilson, R., 174
Wilson, W. J., 128
wind, solar, 61–2
Wollaston, W. H., 16

X-ray astronomy, 92–3, 99